CRISIS COMMUNICATION
CASE STUDIES IN HEALTHCARE IMAGE RESTORATION

RICHARD L. JOHNSON, MA
FOREWORD BY ROBERT E. BROWN, PHD

a division of

Crisis Communication: Case Studies in Healthcare Image Restoration
by Richard L. Johnson, MA

Published by HCPro, Inc. Copyright ©2006 HCPro, Inc.

All rights reserved. Printed in the United States of America.   5 4 3 2 1

ISBN 1-57839-860-6

No part of this publication may be reproduced, in any form or by any means, without prior written consent of HCPro, Inc., or the Copyright Clearance Center (978/750-8400). Please notify us immediately if you have received an unauthorized copy.

HCPro, Inc., provides information resources for the healthcare industry.

HCPro, Inc., is not affiliated in any way with the Joint Commission on Accreditation of Healthcare Organizations, which owns the JCAHO trademark.

Robert E. Brown, Ph.D., Reviewer	Jean St. Pierre, Director of Operations
Matthew Cann, Group Publisher	Darren Kelly, Production Coordinator
Doug Ponte, Cover Designer	Leah Tracosas, Copyeditor
Jackie Singer, Graphic Artist	Paul Singer, Layout Artist

Advice given is general. Readers should consult professional counsel for specific legal, ethical, or clinical questions.

Arrangements can be made for quantity discounts. For more information, contact

HCPro, Inc.
P.O. Box 1168
Marblehead, MA 01945
Telephone: 800/650-6787 or 781/639-1872
Fax: 781/639-2982
E-mail: *customerservice@hcpro.com*

Visit HCPro at its World Wide Web sites:
www.healthleadersmedia.com,
www.hcpro.com, and *www.hcmarketplace.com*

9/2006
20953

CONTENTS

Acknowledgements vi
About the author vii
Foreword: The Age of Crisis viii

Chapter 1: Introduction 1
    The importance of image 3
    The rhetoric crisis 5
    Crisis defined 6
    Methodology 7
    Additional readings 9
    Endnotes 10

Chapter 2: Duke University Health System
hospitals wash surgical tools in hydraulic fluid 11
    About Duke University Health System 13
    Background of the controversy 14
    Duke's stakeholders 20
    Analysis of efforts to restore Duke University
    Health System's public image 21
    Conclusion 29
    Implications for action 30
    Endnotes 32

Contents

**Chapter 3: Children's Hospital deals with
child molestation and pornography accusations** 35
    About Rady Children's Hospital 37
    Background of the controversy 38
    Children's stakeholders 42
    Analysis of efforts to restore Children's public image 43
    Conclusion 49
    Implications for action 49
    Endnotes 52

**Chapter 4: Kaiser Permanente's Northern California kidney transplant
program falls under attack** 53
    About Kaiser Permanente 55
    Background of the controversy 56
    Kaiser's stakeholders 60
    Analysis of efforts to restore Kaiser's public image 61
    Conclusion 72
    Implications for action 73
    Endnotes 76

**Chapter 5: Loyola University Medical Center
responds after infant abduction** 79
    About Loyola University Medical Center 81
    Background of the controversy 82
    Loyola University Medical Center's stakeholders 85
    Analysis of efforts to restore Loyola University Medical Center's
    public image 86

Conclusion 88
Implications for action 88
Endnotes 90

**Chapter 6: Hacker accesses patient information
at University of Washington Medical Center** 91
   About the University of Washington Medical Center 93
   Background of the controversy 94
   University of Washington Medical Center's stakeholders 97
   Analysis of efforts to restore the University
   of Washington Medical Center's public image 97
   Conclusion 100
   Implications for action 101
   Endnotes 102

Acknowledgements

I have benefited from many supporters. Here are just a few to whom I owe my gratitude. I would like thank Dr. Robert Brown for his considerable insight and guidance. It gives me great pleasure that he would write the foreword for this book. I am also indebted to Matthew Cann and Robert Stuart for supporting my work at HCPro. My HealthLeaders Media colleagues not only offered suggestions for case studies but also covered my duties at our magazine and Web site so I could research and write this book. Lastly, my family—Cindy, Trevor, and Annabelle—continually tolerated with good humor my absentmindedness while I pondered too much about healthcare crisis communication.

About the author

Richard L. Johnson is a writer for HealthLeaders Media, a division of HCPro. His articles can be found in *HealthLeaders* magazine and at *www.healthleadersmedia.com*. Prior to returning to healthcare journalism, Johnson worked in corporate communications for a leading financial services company. While pursuing his master's in communication at Suffolk University, he concentrated his studies in organizational communication and development. He also has extensive experience in executive communication. At HCPro, he has written articles and helped authors develop manuscripts. Johnson is a member of the Association of Health Care Journalists. He may be reached via e-mail at *rjohnson@healthleadersmedia.com*.

Foreword

The Age of Crisis

In 1948, the poet W. H. Auden published a collection of poems that captured the postwar mood. He called it *The Age of Anxiety*. Today, a title equally in touch with the zeitgeist might be "The Age of Crisis."

We live and die in the wake of such events as 9-11, the emergence of HIV/AIDS, and the devastation of Hurricane Katrina. We anticipate plague-like pandemics from SARS to bird flu. We stock our cellars, wondering how to survive in a world where risk borders inevitability. Meanwhile, wary organizations write crisis-management plans and hire leaders with compelling track records in crises.

But while we endure this era of dread, we also benefit from lessons learned about the complex phenomenon of crisis. If this isn't the golden age of crisis scholarship, it may be the most prolific. In recent decades, we have learned much about crises of all kinds, including those affecting healthcare organizations. Scholars and practitioners of social science, philosophy, and mathematics have multiplied our understanding and heightened our awareness. If knowledge is power, then organizations have become better equipped to manage in the age of crisis.

The recent history of organizations and nations has generated a rich vein of cases. Analysts now have a large and growing sample frame for observation, interpretation, and theory building. In addition to political and corporate scandals (Watergate, Monica-gate, Enron), health-related crises have hit the headlines (or have been kept out of the news), providing much material for empirical research. The mass public has learned not only that bad things happen to bad organizations like Enron but also that bad things can happen to otherwise good organizations. Not long ago, a prestigious Boston-area children's hospital with a global reputation for excellence suffered a major reputational loss when hospital management system failures led to the deaths of small children who were left unattended. Perhaps above all, we have learned that there is no single type of crisis and no one-size-fits-all crisis management strategy.

First generation crisis communication

Twenty years ago, textbooks in communication tended to focus on iconic cases. Foremost among these is the case of Tylenol, in which a number of individuals died after ingesting Tylenol capsules purchased from a Chicago drugstore. A myth of excellence grew up around James Burke, the CEO of Johnson & Johnson, the parent company of the makers of Tylenol.

Burke drew praise for his open stance: an upfront, rapid, truthful, bold, and apparently candid response to the crisis. The first generation of scholarship about crisis, including health-related crisis, was associated with big company names and a cast of heroes and villains.

There was the drunken captain who ran aground the Exxon Valdez and spilled oil in pristine Alaska waters. He was paired with the Exxon's CEO, the clueless Lawrence G. Rawl. There was President John Kennedy's gutsy brinkmanship that scuttled the Cuban Missile Crisis. There was the NASA Challenger and the

Foreword

healing rhetoric of President Ronald Reagan's speech to the families and the nation. More recently, there was the September 11 attack of the World Trade Center and the Churchillesque visibility of New York City's Mayor Rudolf Giuliani.

In first-generation crisis communication books such as Laurence Barton's *Crisis in Organizations* (1993), the lessons were large and the patterns relatively few. And although the second generation of scholarship has discovered far more complex patterns, many of the first generation lessons remain axiomatic. For example, organizations should do everything to discover potential crises before they detonate into life-shattering, reputation-altering events. In the event of a crisis, management should move quickly, but not precipitously, to get the facts. There should be a single spokesperson, not a cacophony of messages. Crisis affects the entire organization and its stakeholder constituencies, so these constituencies must be quickly and clearly informed. Above all, management should not lie or, even worse, attempt to cover up, as was the unfortunate case with all the "gates" of disgraced presidents from Nixon to Clinton.

After September 11, Enron's shadowy, unethical culture and splashy, tragic debacle made transparency a watchword. "No comment" was out; accountability was in. But long before Enron, the cynical spin known as "damage control" had gone bankrupt. The damage done at Union Carbide's facility at Bhopal, India, in 1985—nearly 2,000 lives lost—could not be controlled. Health-related organizations and others have come to believe that crisis is not a matter of "if" but "when."

In the 1990s, companies got bullish on crisis management plans (CMPs). But even these had their limitations. For as generals have long known, wars (and other crises) make planning necessary but plans useless. In the Internet age, the world is an unexpected whirl of contingencies. As the management theorist Ira Mitroff pointed out with the subtitle to his book, *Crisis Leadership: Planning for the Unthinkable* (2004), even "prepared" organizations could not inoculate themselves against crises. Mitroff notes that, prior to September 11, not a single major company's CMP tabbed terrorism as a high-likelihood risk. Like most of us, the smartest managers in the room were taken by surprise.

New approaches to crises

Crisis communication entered a mature phase in the decade that followed the Valdez and preceded the World Trade Center attacks. As experts studied the mounting number and types of crises, they came to identify whole new levels of complexity. Crisis couldn't be bottled; icons no longer sufficed. As a result, a new generation of post-9-11 scholarship has blossomed. New theory-based approaches have been developed and tested, including Situational Crisis Communication Theory, developed in recent years by public relations scholar Timothy Coombs. Increasingly, business management scholars and consultants discovered fresh insights about crisis in the scholarship of speech communication, rhetoric, and public relations. In the new era of crisis management, communication researchers tackled crises and produced and tested complex, multilevel remedial strategies. With experience in the political and corporate arenas, analyzing such image-shattering crises as Exxon's and President Bill Clinton's, rhetoric scholar W. L. Benoit developed a theory of "image repair" (or restoration). Richard L. Johnson's book, published by HealthLeaders Media, is a new-generation product of Benoit's theory.

In the period following the emergence of HIV/AIDS, crisis communication scholarship has been dominated by rhetorically based theory that dates back to Aristotle's *Ars Rhetorica* in the Greece of fourth century B.C.

FOREWORD

That book, the classic foundation of public relations and crisis communications, is the study of how speakers systematically analyze their audiences and situations to craft persuasive messages. Using examples from poets, dramatists, generals, and politicians, Aristotle created a taxonomy of audience types, situations, strategic arguments, methods of organization, and a lexicon of eloquent language. Aristotle is the true father of the new wave of rhetorical and narrative crisis communication.

Following Benoit, Johnson places rhetoric and storytelling front and center in the crisis management endeavor. For the benefit of managers charged with crisis management responsibility and reputational protection, Johnson's book aims not only at analyzing cases but also at deriving principles and strategic management decisions.

Today, management science has rediscovered the ancient art of rhetoric and turned to its modern institutional expression: public relations. In this tradition, Johnson's case studies make a valuable contribution to the literature of organizational crisis. This readable volume succeeds in integrating the domains of scholarship and practice. The cases are certain to provide thoughtful support for decision-makers and rich material for students and scholars of organizational crises.

> **Robert E. Brown, Ph.D.,** is Professor of Communications at Salem State College in Massachusetts. He is a former speechwriter for Atlantic Richfield (BP), W. R. Grace & Co., and was a marketing manager at Coopers & Lybrand (Price-Coopers). Dr. Brown serves on the editorial board of the *Public Relations Review*, a scholarly journal to which he is also a frequent contributor. He lectures on crisis communication at the Harvard University Extension School and Emerson College. His memberships include the National Communication Association, Public Relations Society of America, and the International Academy of Business Disciplines.

Introduction

C H A P T E R   1

Introduction

The importance of image

Whether you work for a prominent top-tier medical institution or a small rural hospital, your organization is looked to as a savior by those in your community. Your stakeholders have one expectation of your organization and those working in it: perfection. This might seem unrealistic and unfair to healthcare leaders, who face many of the same challenges as managers in other industries—complex and changing rules, intense regulation, disruptive employees, defective products, increased volume, decreased reimbursement, mounting competition, and intense public scrutiny.

Elvis Presley once said, "The image is one thing, and the human being is another. It's very hard to live up to an image."[1] With a legion of fans calling

CHAPTER ONE

him "the king," he should know. Even though it might seem daunting—if not impossible—to live up to the high expectations the public has for your healthcare organization, savvy managers realize that a strong public image is a bankable commodity.

How long do you suppose it takes a PR manager to spring into action to alert the media after learning that her hospital has cracked into the top 10 on *U.S. News & World Report's* annual best hospitals list?

So it is safe to assume that public image is important to healthcare organizations. And when a healthcare organization's reputation is under attack, it stands to reason that the organization will attempt to implement strategies to reduce the effectiveness of the attack and maintain its good public image. A recent study by the Public Relations Society of America's Health Academy supports this notion. A survey of healthcare executives found that two-thirds rated reputation as "critical" or "very important" to the success of their organizations, and 86% reported that they would expect negative events to affect their reputations and bottom lines adversely.[2]

If we can agree that public image is important to healthcare organizations, then we can also assume that organizations engage in a variety of activities to influence the public's perception of them. Every day healthcare executives, clinicians, and support staff attempt to construct, reconstruct, and maintain the public image of their organizations. They do this in ways that are strategic and in ways that are perhaps unintentional. We observe these attempts firsthand—when the large health system spends $40 million for an advertising campaign, when the hospital CEO gives a speech at major industry convention, when a kindly physician takes the time to carefully explain a procedure, and when a security guard volunteers to help a lost family member find a patient's room.

The rhetoric crisis

When a healthcare organization experiences a crisis event, the words and actions of its officials take on greater significance. Not responding to the crisis in hopes that the issue will go away allows the organization's critics to influence its stakeholders through media outlets. As Keith Michael Hearit notes, an organization's silence can be "tantamount to an admission of guilt, given our media-driven constructions of social reality in contemporary society."[3] Enter the PR practitioner, whose role is to use the power of rhetoric to increase the probability of positive public perceptions about the organization's credibility and image.[4]

Thus, I began thinking about how healthcare organizations respond rhetorically to crisis events. For an understanding of a rhetorical approach to crisis communication, I turn to Dan P. Millar and Robert L. Heath: "A rhetorical approach to crisis asks the following question: What needs to be said before, during, and after a crisis?"[5]

If we apply this question to the healthcare industry, where public trust is one important measure of a healthcare institution's success, we can imagine that the answers could be invaluable to executives, PR practitioners, and clinicians alike. Even though there are numerous case studies of crisis communication involving healthcare organizations, I could not find a book of case studies that applied theoretical framework to the study of healthcare crisis communication. This is an important distinction. Case studies that are grounded in theory, in my opinion, hold the writer to a higher standard of accountability to help the reader by providing prose that is more detailed, exhaustive, and prescriptive.

Chapter One

Crisis defined

A review of communication literature produces a number of different definitions of crisis. Here is a sampling from notable communication scholars:

Seeger, Sellnow, and Ulmer define an organizational crisis as "a specific, unexpected, and nonroutine event or series of events that create high levels of uncertainty and threaten or are perceived to threaten an organization's high-priority goals."[6]

According to Fearn-Banks's definition, "a crisis is a major occurrence with a potentially negative outcome affecting the organization, company, or industry, as well as its publics, products, services, or good name. A crisis interrupts normal business transactions and can sometimes threaten the existence of the organization."[7]

Caywood and Stocker define a management crisis as "an immediate unexpected event or action that threatens the lives of stakeholders and the ability of the organization to survive."[8]

Finally, in an interesting opinion that is significant for the rhetoric of crisis, Hearit and Courtright note, "[W]e argue for a conceptualization that locates all crisis management as crisis communication management, because crises are both created and resolved communicatively. Said another way, we assert that communication, rather than being one variable in the crisis management mix, actually constitutes the nature and experience of the crisis itself."[9]

Methodology

In the following chapters of this book, I analyze five case studies in which healthcare organizations in the midst of crisis events attempted to influence public opinion. I do so using William L. Benoit's theory of Image Restoration Strategies as a rhetorical framework. This theory includes a typology of image restoration strategies that "are organized into five broad categories, three of which have variants or subcategories: denial, evading responsibility, reducing offensiveness, corrective action, and mortification."[10] According to Benoit, an attack on an organization's image has two components: An act has to be considered offensive by stakeholders, and the organization accused has to be held responsible for that act.[11]

I use image restoration as the theoretical framework for these case studies because it allows me to address the concerns healthcare organizations have about public image, and it permits me to give preference to the rhetoric of crisis events. Further, this theory is extraordinarily adaptable to a variety of cases, and its acceptance is growing within communication. I hope that this approach will allow the reader to digest these case studies with ease and compare and contrast the cases included in this book. Moreover, image restoration—and, in fact, this book—is intended to be not only descriptive but also prescriptive. The reader should actively critique the choices made by the organizations mentioned in this book and consider carefully how the reader's organization might respond to similar situations and attacks.

Each case study in this book is organized similarly. First, I provide a brief overview of the crisis situation and information about the organization being studied. Next, I describe the background of the controversy. I list the organization's salient stakeholders. Then I offer a detailed analysis of the organization's efforts to restore its

Chapter One

Benoit's image restoration strategies

Denial
- Simple denial
- Shifting the blame

Evading of responsibility
- Provocation
- Defeasibility
- Accident
- Good intentions

Reducing offensiveness of the event
- Bolstering
- Minimization
- Differentiation
- Transcendence
- Attacking accuser
- Compensation

Corrective action

Mortification

public image. I give an analysis of the overall communication strategy. Finally, I present implications for action, should the reader's organization ever face similar attacks.

It is my intention to provide readers with one framework to analyze crisis communication, but I also hope my work will encourage others to publish more analyses of healthcare communication case studies. Given the vast number of communicative events in the healthcare industry—many of which have tremendous impacts on society—there is an opportunity to develop a wealth of literature that can benefit healthcare executives, PR practitioners, clinicians, and communication researchers.

Additional readings

There are many methods of crisis communication analysis. I strongly encourage readers to apply these other works to their crisis communication planning efforts. If you find Benoit's work useful, you may be interested in Hearit's book related to apologia, *Crisis Management by Apology: Corporate Responses to Allegations of Wrongdoing* (2005). In addition, W. Timothy Coombs is widely published and has advanced the theory of Situational Crisis Communication, which "suggests that an organization's past crises history affects the reputational threat posed by current crisis when that crisis results from intentional acts by the organization" (Coombs, W. T. [2004]. Impacts of past crisis on current crisis communications: insights from situational crisis communication theory. *Journal of Business Communication*, 41 [3], 265–289.). Finally, Millar and Heath edited a fine collection by communication scholars titled *Responding to Crisis: A Rhetorical Approach to Crisis Communication* (2004).

CHAPTER ONE

Endnotes

1. Brian McTavish, "Still Dead After All These Years," *The Kansas City Star* (August 16, 1997): E1.

2. Public Relations Society of America's Health Academy, "Too Late to Triage: Is Health Care's Reputation Headed for the Critical List?" Public Relations Society of America's Health Academy, *www.healthacademy.prsa.org/pdf/CEO_Survey_Press_Release.pdf* (accessed July 26, 2006).

3. Keith Michael Hearit, "Apologies and Public Relations Crises at Chrysler, Toshiba, and Volvo," *Public Relations Review* 20, no. 2 (1994): 114.

4. Robert E. Brown, "A Matter of Chance: The Emergence of Probability and the Rise of Public Relations," *Public Relations Review* 29 (2003): 387.

5. Robert L. Heath and Dan P. Millar, "A Rhetorical Approach to Crisis Communication: Management, Communication Processes, and Strategic Responses," in Millar, D. P., and Heath, R. L. (eds.), *Responding to Crisis: A Rhetorical Approach to Crisis Communication* 2004. Lawrence Erlbaum Associates, Publishers.

6. Matthew W. Seeger, Timothy L. Sellnow, and Robert R. Ulmer, "Communication, Organization, and Crisis," *Communication Yearbook* 21 (1998): 233.

7. Kathleen Fearn-Banks, *Crisis Communications: A Casebook Approach*, 2nd ed. (New Jersey: Lawrence Erlbaum Associates, Publishers, 2002), 2.

8. Clarke L. Caywood and Kurt Stocker, "The Ultimate Crisis Plan," in Gottschal, J. A. (ed.), *Crisis Response: Inside Stories on Managing Image Under Siege*. Visible Ink.

9. Keith Michael Hearit and Jeffrey L. Courtright, "A Symbolic Approach to Crisis Management: Sears' Defense of Its Auto Repair Policies," in Millar, D. P., and Heath, R. L. (eds.), *Responding to Crisis: A Rhetorical Approach to Crisis Communication* 2004. Lawrence Erlbaum Associates, Publishers.

10. William Benoit, *Accounts, Excuses, and Apologies: A Theory of Image Restoration Strategies* (New York: State University Press, 1995), 75.

11. William Benoit, *Accounts, Excuses, and Apologies: A Theory of Image Restoration Strategies* (New York: State University Press, 1995), 71.

2

Duke University Health System hospitals wash surgical tools in hydraulic fluid

Two Duke University Health System hospitals notified 3,800 patients that the surgical instruments used during their operations may not have been properly cleaned. Through a series of mix-ups, staff at Duke Health Raleigh Hospital and Durham Regional Hospital substituted used hydraulic fluid, which had been drained from an elevator, for surgical equipment detergent. Critics attacked Duke for not swiftly correcting the error or quickly sharing vital information about the contents of the fluid with affected patients, some of whom claimed to become ill after their surgeries.

CHAPTER 2

Duke University Health System hospitals wash surgical tools in hydraulic fluid

About Duke University Health System

Duke University Health System is a nonprofit academic healthcare system headquartered in Durham, NC. Its hub is Duke University Hospital, which has been named as one of the top 10 U.S. hospitals in the annual *U.S. News & World Report* list of best hospitals for the 17th year in a row. Included in the health system are three hospitals (Duke Hospital, Durham Regional Hospital, and Duke Health Raleigh Hospital); ambulatory surgery centers; primary and specialty care clinics; and home care, hospice, skilled nursing care, and wellness centers. For fiscal year 2004, Duke University Health System reported an operating income of $75.6 million from approximately $1.5 billion of revenue. This was its most profitable year on record.

Chapter Two

Background of the controversy

Duke University Health System reached out to patients and the media in January 2005 to report a regrettable error. Staff at two Duke University hospitals—Duke Health Raleigh and Durham Regional—unwittingly washed surgical instruments with hydraulic fluid, thinking it was detergent. The error was repeated from early November until late December 2004, and Duke estimates that 3,800 patients might have been operated on with the improperly cleaned tools.[1]

In initial news coverage, Duke officials offered assurance that patient care was not compromised and that patients who may have been exposed did not need to be screened for bloodborne diseases. They said the surgical tools had been sterilized in a separate step after being improperly washed.

Little threat to patients

An expert on sterilization from the University of North Carolina backed up Duke's assertion. "Steam sterilization is an extraordinarily robust process," said Dr. William Rutala. "You can do a lot of things wrong and still have a sterile instrument."[1]

In a story by the *Herald-Sun*, Wayne Thomann, a Duke occupational safety specialist, said the hydraulic fluid was relatively harmless and that the U.S. Environmental Protection Agency did not consider it hazardous.[2] Nonetheless, Duke voluntarily reported this incident. In letters to the 3,800 patients, Duke urged them to contact the health system if they noticed signs of infection. "If it happened to me, I'd rather have someone tell me than not," said James P. Knight, chief executive of Duke Health Raleigh Hospital.[1] Duke said it was monitoring surgical infection rates and had not seen a significant increase at the hospitals.

The hydraulic fluid had been drained from an elevator at Duke Health Raleigh Hospital by a worker from a company called Automatic Elevator. The fluid was put into about a dozen empty 15-gallon drums labeled as Mon-Klenz, a surgical equipment detergent. These drums were left at the hospital and later mistaken for surplus cleaner. A representative from Cardinal Health, the detergent supplier, brought the drums back to its warehouse, also believing it was detergent.[2] In November and December 2004, when the hospitals ordered more detergent, Cardinal Health delivered the drums of used hydraulic fluid marked Mon-Klenz to Duke Health Raleigh, Durham Regional, and Wake Forest University Baptist Medical Center in Winston-Salem.[3] According to Duke officials, staff did not initially notice the error because Mon-Klenz and hydraulic fluid are both odorless and have a similar amber color.

The day after the story about the mix-up broke, a follow-up article suggested that injury to patients was unlikely. Because the sterilization process was so effective, experts interviewed by *Herald-Sun* reporter Jim Shamp said the hospital probably did not harm patients. "This is clearly a goof, probably a negligent act, but I don't see where there's much likelihood of an injury to patients," said Bill Faison, a Durham malpractice attorney. "It's a bad thing to use a hydraulic fluid instead of the detergent they intended, but if patients were going to have a reaction, odds are it would have happened immediately."[4]

A government report changes the story

In these initial reports, Duke appeared at least partially responsible for the problem, but the consequences of the problem on its stakeholders do not seem to be very great. Although some criticized Duke for the error, others praised the organization's leadership for disclosing the problem and taking corrective action.[3] But many more critics emerged in June 2005 when the *News & Observer* released details of a Centers

CHAPTER TWO

for Medicare & Medicaid Services report about the incident.[3] Investigators for the agency determined that "an immediate jeopardy to patients" existed at Durham Regional and Duke Health Raleigh hospitals and cited both facilities for noncompliance with Medicare regulations from November 3, 2004, to January 5, 2005.[4]

"Administrative staff failed to heed the multiple complaints of staff sterilizing and using the instruments, thus delaying the discovery of the error and needlessly exposing patients to these instruments over a longer time period," stated the CMS report.[3]

Duke's image was attacked by patients, their families, and lawyers who claimed that Duke officials refused to tell them the contents of the hydraulic fluid. Some patients reported illnesses stemming from their surgical procedures and expressed frustration with Duke. "What we really want is a response from Duke to prove their assertions that there was very little risk to the patients," said attorney Thomas Henson. "I mean, patients are hanging out there with problems, and Duke won't give an answer to us."[5]

"We put our faith in Duke and in one way, they failed us," said patient Carol Svec, who reported a slow recovery after surgery at Duke Health Raleigh. "It seems to me that a good company with a good reputation should be bending over backward for us."[6]

Duke responds to critics

In response to this criticism, Dr. Victor J. Dzau, chancellor for health affairs at Duke University and president and CEO of the health system, said that Duke had hired an outside lab to analyze the hydraulic fluid and would make the results known.[7] He also downplayed the importance of the contents of the hydraulic fluid.

He said it was more important to consider what trace elements from the fluid would be left on surgical instruments after being washed with a mix of water and hydraulic fluid and then being sterilized. "But the really important analysis is what's on the [surgical] implements, not what's in the bulk fluid. The patient only comes in contact with the scalpel, the retractor, things like that. So the problem is to take the trace chemicals off the metal implements, because that's what patients would have been exposed to, if anything at all," Dzau said.[8]

Duke also publicized surgical infection rates at the two hospitals for the two months in which patients would have been exposed to the improperly cleaned instruments. Duke Health Raleigh's surgical infection rate was up slightly with a rate of 1.4%. Over a two-year average, its surgical infection rate was 0.44%. At Durham Regional, the surgical infection rate was essentially unchanged. It was 0.99% over the two-month period, and its two-year average was 0.95%.[7]

In an effort to open the lines of communication with affected patients, Duke also set up a telephone hotline, hired a patient liaison officer, created a system to track the health of affected patients, initiated a patient advocate committee, and published a Web site for concerned patients.[9] These changes prompted Svec, the patient who complained about her slow recovery, to change her opinion of Duke. "I'm thrilled that their level of response is sincere," she said. "The people who are trying to deal with this problem are very aware now of what these patients need."[9]

Lastly, Duke enlisted two groups of scientists, who were unaffiliated with Duke, to conduct experiments and analyze whether the surgical tools that Duke Health Raleigh and Durham Regional used could have harmed patients. These analyses were detailed in two letters to patients dated June 20, 2005, and June 27, 2005. In the first experiment, the Statewide Program in Infection Control and Epidemiology at the University of North Carolina School of Medicine found that "replacing cleaning detergent with hydraulic fluid did not alter the effectiveness of the sterilization process as high numbers of clinically relevant bacteria and standard test spores (relatively resistant to the sterilization process) were completely inactivated."[10]

Good news for patients

In the second letter from Duke, officials furthered their claim that patients were unharmed by the error. RTI International conducted an analysis "to determine how much hydraulic fluid remained on surgical instruments after they had been rinsed in very hot water and then sterilized, as was the case in the incident at Durham Regional Hospital [and Duke Health Raleigh Hospital]."[11] The researchers "concluded that the residual amount of fluid on the instruments was very small, approximately 0.08 milligrams per instrument, on average. This is the equivalent of 0.002 of a drop."[11] The letter, signed by Dzau, said that the findings were good news for patients worried about the effects of improperly cleaned surgical tools. It also said that Duke was slow to release this information because the independent studies were very extensive and because the maker of the hydraulic fluid, ExxonMobil, was slow to respond to Duke's request for information. "This information was not provided to Duke until June 7 because the company considers it proprietary—in layman's terms, a trade secret. ExxonMobil provided the information under the condition that it would be made available only to the scientists conducting the analyses."[11]

In a newspaper report about the letters to patients, a spokesman reiterated Duke's claim that it took months for ExxonMobil to provide Duke with the chemical composition of its hydraulic fluid and would do so only conditionally. "Our preference would be to provide our patients with the full list of chemical ingredients in the hydraulic fluid," said spokesman Jeff Molter.[12] ExxonMobil responded with a press release saying that Duke was not up front when it requested information about its hydraulic fluid. "Had Duke informed ExxonMobil of the nature of the issue, ExxonMobil's initial response would have been equally immediate and unrestricted," the company said. "Duke did not indicate that patients may have been exposed to used hydraulic fluid, nor did they ask for an immediate response."[13]

Molter said that Duke believes it communicated clearly with ExxonMobil. On June 28, 2005, ExxonMobil told Duke that it was free to share information about the contents of the hydraulic fluid with all patients without the need of a confidentiality agreement.[13]

During the summer of 2005, lawsuits were filed against the Cardinal Health and Automatic Elevator. In court, lawyers for Duke University Health System agreed to provide an attorney representing seven patients with a sample of the used hydraulic fluid along with information about its chemical composition.[14] It was not until July 2006 that a patient sued Duke University Health System. In a lawsuit, Bennie W. Holland Jr. said that he developed an infection after back surgery at Duke Health Raleigh Hospital.[15]

"The evidence we've seen to date suggests this low exposure was not harmful to our patients," Duke said in a prepared statement. "We regret this incident occurred, but stand by the results of independent studies and our own analysis."[15]

Chapter Two

Duke's stakeholders

The following key players influenced Duke's public perception during the crisis event:

- Senior leaders
- Duke's news office staff
- Patients and their families
- Patient's lawyers
- Federal regulators
- State regulators
- ExxonMobil
- Cardinal Health
- Media

Through a variety of methods, Duke University Health System interacted with these key stakeholders. It made senior officials available to the media, created a press kit about the issue, sent letters to affected patients, set up a hotline, and published a Web site. In addition, Duke's chancellor for health affairs sent Duke physicians and staff a letter detailing the organization's position on the matter. Throughout its communication efforts, Duke stood firm that the outside experts it hired found that its surgical sterilization process had not been compromised and that patients had not been harmed due to exposure to hydraulic fluid.

Analysis of efforts to restore Duke University Health System's public image

In initial news reports, Duke acknowledged that a problem occurred at two of its hospitals, but the immediate offensiveness of the act did not appear to be substantial. This case study is a particularly good example of the power of framing a story for the media. Because Duke discovered the hydraulic fluid mix-up, it had the opportunity to frame the situation as an unfortunate mistake that could not harm patients. Indeed, sources quoted in the news from outside of the Duke University Health System substantiated this position. It was not until four months after Duke disclosed the problem that the majority of news stories came out criticizing Duke. The report from CMS—in which investigators said that Duke administrators put patients in "immediate jeopardy" and was slow to correct the problem—set in motion the second, more powerful wave of rhetorical attacks against Duke's public image. In the remainder of this section, I show how Duke University Health System sought to defend its public image using these image restoration strategies: evading responsibility by showing good intentions, denial by shifting blame, reducing the offensiveness of the act by minimization, and corrective action.

Good intentions

As noted above, Duke's first communication efforts about the hydraulic fluid show the system in a somewhat positive light. January 7, 2005, stories in the *Herald-Sun* and the *News & Observer* point out that Duke sent letters to affected patients. "If it happened to me, I'd rather have someone tell me than not," said Duke's Knight.[1] His colleague, David McQuaid, Durham Regional's chief executive, said that Duke's "primary concern has been to communicate to our patients."[2] Duke officials said an internal investigation found the problem to be a simple mistake and not a deliberate attempt to cause trouble. In addition, letters to patients were sent within days of discovering the problem and encouraged patients to see their physicians if they experienced pain, redness, fever, or other signs of infection.

When Duke found its error, it responded quickly and appropriately. This displayed a genuine concern for patients and for acting ethically. In fact, in one of these initial stories, Faison, the Durham malpractice attorney, praised Duke for communicating to patients: "Several times a year, I find in cases I review that healthcare providers either flat-out falsify a record, or they'll be substantially misleading in the record. So I think in this case I'm encouraged by the fact Duke is being forthright, giving patients an opportunity to see to their own care."[4]

When federal investigators reported that Duke put patients in jeopardy because administrators "failed to heed" complaints by staff,[3] they invited more attacks on Duke University Health System's public image. Critics' chief attack was that the health system ignored patients' reports of complications by not providing the chemical contents of the hydraulic fluid. For example, when Shelley Bassett complained to Dr. Powell G. Fox, Duke Health Raleigh's chief medical officer, that she suffered from aches, nausea, and vomiting, she said that he was dismissive: "He said it doesn't sound to me that it has anything to do with the oil, and that you probably just have the flu. I felt very patronized. If he really truly was there to take people's calls who were having issues, then he should follow through and make sure people are doing better. He should have said we will check you out."[3] Many other patients made similar comments relating to Duke's good intentions, saying that they suffered great discomfort after their surgeries and wanted to know what specifically they were exposed to.

Thomas Henson Sr., a lawyer for eight Duke patients, suggested that Duke was attempting to hide information from the public: "If Duke doesn't have anything to hide, why are they hiding it? Give me the scientific evidence that says these people are not at risk. Tell me how people can go to bed at night and not worry."[3]

By mid-June 2005, Duke's Dzau responded to these attacks with statements to the press and a memo to Duke University Health System's physicians and staff. In these statements, Dzau said Duke had sent samples of the used hydraulic fluid for analysis and planned to share results with patients but that this was a complex process that could take weeks. "We want to give people the message that we care about our patients," he said in the *News & Observer*.[7]

In his memo, Dzau acknowledged that staff and their patients might be concerned by attacks on Duke's reputation in the media: "I feel badly that any patient suffers post-operative complications, but I know that there is always some risk of a poor outcome in any surgery, wherever a procedure is performed, even under the best of circumstances. I want to assure you that the health and welfare of our patients will always be our top priority, and we have given the patients who have contacted us the best information we have had at the time."[16]

The two letters sent to patients, dated June 20 and 27, 2005, also expressed Duke's good intentions, with similar language to the letter to employees noted above.[10, 11]

Shifting the blame

Duke University Health System attempted to shift some of the blame for two separate incidents on other companies. In the first incident, it blamed Automatic Elevator for improperly labeling and securing the used hydraulic fluid its worker drained from a Duke elevator. It also blamed Cardinal Health for restocking the hydraulic fluid as detergent and delivering it to Duke. In the second incident, Duke attempted to shift some of the blame to ExxonMobil for not quickly providing the chemical composition of its hydraulic fluid.

In a document titled *Updated Summary of Response to Hydraulic Fluid Incident* published on a Duke University Health System Web site, the health system claimed that a worker from Automatic Elevator serviced a Duke hospital elevator, drained hydraulic fluid into containers labeled as surgical detergent, did not properly change the label, and did not securely store the containers. The document then said Cardinal Health, the company that provided Duke with surgical detergent, picked up the containers and restocked them. Later, Cardinal Health delivered these containers back to Durham Regional and Duke Health Raleigh hospitals.[17]

It cannot be disputed that this problem began because a worker from Automatic Elevator carelessly emptied hydraulic fluid into detergent drums. In addition, Cardinal Health delivered these drums to Duke hospitals. However, press accounts suggest that Duke University Health System had opportunities to catch this error.

A Duke spokeswoman pointed out that the detergent and hydraulic fluid are similar in appearance. She said Duke staff did not detect the problem because some oiliness of the instruments is common.[2] However, Wake Forest University Baptist Medical Center also received the hydraulic fluid from Cardinal Health, but its staff caught the problem before patients were exposed.[3] This suggests that staff at Duke's hospitals also could have identified and corrected the errors made by Automatic Elevator and Cardinal Health.

Furthermore, a spokesman for Cardinal Health told the press that a Duke employee called the company and requested that it pick up the containers.[4] Also, although the drums were labeled surgical detergent, they were not sealed properly. But Duke workers "accepted them without question," said federal inspectors.[3] The regulators also said that Duke staff complained about greasy surgical tools and attempted to fix the washers but did not check the detergent until late December, even though instructions for the washers suggest checking the detergent if instruments come out unclean.[3] Lastly, despite these complaints, inspectors said that "no surgeries or invasive procedures were canceled based on these complaints."[3]

Duke's attempts to shift blame to Automatic Elevator and Cardinal Health were somewhat successful. All of the organizations involved had a part to play in the incident, and lawsuits against Automatic Elevator and Cardinal Health suggest that patients perceive them to be accountable. However, the public places a higher standard on healthcare organizations to ensure their well-being. As Gail Shulby, patient safety officer for Duke University Hospital, pointed out, "Even if contractors make the mistake, the hospital gets the blame in the eyes of the patient."[18]

In letters to patients and statements to the press, Duke officials gave two reasons for taking several months to provide patients with information about the contents of the hydraulic fluid and tests conducted by outside experts. First, Duke officials said the experiments they had conducted were sophisticated and extensive. It could have received faster results with less valid studies, but Duke said it wanted to provide patients with reliable information.[11] Second, Duke said it was "hampered by delays in obtaining information about the hydraulic fluid from its manufacturer, ExxonMobil." Duke said it requested information as early as February 2005 but did not receive it until June 7, 2005. In addition, the information about the contents were provided on the condition that Duke would share it only with scientists.[11]

This attempt to shift the blame for the delay puts ExxonMobil in a very bad light. Perhaps Duke should have predicted that ExxonMobil would respond to an attack on its own public image. ExxonMobil provided a statement to the press that attacked Duke University Health System for misleading the company by not sharing important details about the incident or requesting an immediate response.[13] ExxonMobil blamed Duke for the delay: "Had Duke informed ExxonMobil of the nature of the issue, ExxonMobil's initial response would have been equally immediate and unrestricted."[13] Duke's reaction was tempered. A spokesman said Duke was pleased to have the information about the fluid and ExxonMobil's permission to share it with patients.[13]

This attempt to shift partial blame to ExxonMobil was not effective. Clearly, ExxonMobil did not believe it was responsible for the delay, and one could predict that any company would attempt to defend its image from attack. Moreover, by mentioning ExxonMobil in its letters and to the press, Duke introduced another facet of the story for journalists to investigate. This led to additional news coverage about the hydraulic fluid incident that could have been avoided. Perhaps more importantly, Duke had already provided a reasonable excuse for taking so long to provide this information to patients. With a matter as important as patient safety, it makes perfect sense to conduct such scientific experiments in a deliberate fashion. Duke did not need to offer additional information about ExxonMobil.

Minimization

Duke was attacked for two related offensive acts. First, Duke attempted with some success to minimize charges that patients were put in danger because it substituted hydraulic fluid for detergent. Second, Duke attempted to minimize attacks to its image about not responding promptly to patient concerns.

Duke University Health System hospitals wash surgical tools in hydraulic fluid

In the initial newspaper reports, Duke said patient care was not compromised because the instruments that were washed with a combination of hydraulic fluid and water were properly sterilized in a separate step.[1] This is a position Duke held throughout news reports and in its own communication efforts. Duke further pointed out that surgical infection rates for both hospitals were not statistically higher during the period when the improperly cleaned instruments were used and that the hospitals' surgical infection rates were well within the standard 1%–2% seen at similar hospitals.[8]

To substantiate its claim that patients were not harmed because of the fluid mix-up, Duke also commissioned two scientific experiments to be conducted by outside experts. These studies re-created the situation at two of the health system's hospitals to find out whether the surgical instruments used were sterile and whether the used hydraulic fluid present on the instruments could have harmed patients. The first test found that "replacing cleaning detergent with hydraulic fluid did not alter the effectiveness of the sterilization process. . ."[10] The second test by outside experts concluded that based "on this new information, along with a full list of the chemical ingredients in the fluid provided by the manufacturer, ExxonMobil, we can now tell you with even greater certainty that this very low exposure, on the broad surface of an instrument or device, was not harmful to patients."[11]

Faced with attacks that it put patients in jeopardy, Duke University Health System used scientific findings to refute its critics. Coupled with its relatively low surgical infection rate, this evidence is very compelling, and Duke did a fine job of explaining it in scientific and simple terms to its patients, reporters, and its physicians and staff.

Perhaps just as damaging to Duke's public image, several newspaper reports included comments from worried patients and their family members who felt that Duke did not respond to their concerns with a sense of urgency. In just about any crisis situation that has a clinical element, you can bet a reporter will find a distraught patient to interview. For the most part, Duke provided an effective defense for why it did not respond quickly. As outlined above, Duke said that the scientific studies it commissioned were complex and time consuming. Certainly, it is unfortunate that patients have reason to be concerned about their health; however, this situation could have been far worse for Duke and its patients if Duke officials provided information that could not be substantiated. When faced with public scrutiny, there is always pressure to respond, but Duke University Health System showed discipline in its communication approach by only providing information that it could prove accurate.

Corrective action

As with the minimization strategy it used above, Duke implemented the strategy of corrective action to address two attacks: that conditions at Duke hospitals caused hydraulic fluid to be mistaken for detergent and that it did not respond to patient concerns.

First, Duke made a number of changes to ensure that the hydraulic fluid error would not be repeated. CMS accepted corrective action plans from Durham Regional and Duke Health Raleigh,[18] and Duke alerted the media that it took corrective action immediately upon discovering the problem. Duke also put in place new processes for verifying products used to clean and sterilize instruments, established a system to monitor containers, and provided safety training to its staff.[17]

"We're encouraging people to speak up about these things and encouraging our front-line managers to identify things early on that could cause problems," said Karen Frush, chief patient safety officer for the health system. "We're trying to develop a safety culture."[8]

Second, to respond to patients who may have been exposed to improperly cleaned tools, Duke set up a telephone hotline, hired a patient liaison officer, created a system to monitor the health of affected patients, initiated a patient advocate committee, and published a Web site for concerned patients.[9] In addition, it told affected patients that they and their physicians can consult free of charge with two Duke physicians who are experts in environmental medicine.[17] Duke informed patients about these corrective actions through letters and also included the information in its online press kit.

Duke University Health System spent significant resources to implement its corrective action strategy. Overall these efforts were effective. As one previously vocal critic of Duke noted, "I'm thrilled that their level of response is sincere. The people who are trying to deal with this problem are very aware now of what these patients need."[9]

Conclusion

Duke University Health System's attempts to restore its public image were quite successful overall. In the press, Duke officials spoke with one voice, and the majority of their statements appeared to be credible. Duke's strongest efforts were related to its minimization strategy. It created a strong, evidence-based case, minimizing the effectiveness of charges that patients could have been harmed from exposure to hydraulic fluid. Although its communication strategy was deliberate and disciplined, it is unfortunate that patients had to wait many months from the time Duke notified them of the problem until Duke released information about the contents of the hydraulic fluid and the test results from outside experts. If it had been possible for Duke to provide this information sooner, it would have diminished the effectiveness of critics who attacked Duke for not responding urgently to patients'

CHAPTER TWO

concerns. In addition, Duke's rhetorical attack on ExxonMobil backfired; it created additional negative news coverage that should have been anticipated.

Implications for action

This case study shows healthcare executives how their institutions' public image can be damaged by contractor errors. It also highlights how to respond to concerns about medical equipment mishaps and patient exposures. Below are some key lessons to learn from Duke University Health System's experience.

Get your story straight
Duke had the advantage of discovering the perceived offense. This is far preferable to having an outsider find and report an error. Duke quickly investigated the issue, corrected the problem, formulated its strategy, and notified patients and the media. When reviewing the literature in the press and from Duke's news office, statements from Duke officials were consistent.

Find support for your position
Duke held firm that the hydraulic fluid mix-up did not harm patients. Officials not only stated beliefs but they made efforts to prove them. Oftentimes, your statements to the media are no more credible than your attackers'. Be prepared to back up your claims with hard evidence whenever possible. Better yet, get outside experts to do it for you.

Communicate directly to stakeholders

Your statements to journalists will be altered or cut entirely from news stories. To create a compelling counter-narrative to attacks on your public image, take your case directly to your patients and staff. Duke provided journalists with a lot of material, but it also published its position on the Web and in letters to patients and staff.

Get creative

Don't just communicate with words. Think about how images could effectively communicate your image restoration strategies. On Duke's "Hydraulic Fluid Facts" Web site (*hydraulicfluidfacts.dukehealth.org*), it used images to give users a tour of its sterilization area. Also, Duke included photos of its physicians and showed a side-by-side comparison of hydraulic fluid and surgical detergent.

Respond promptly

Cynics abound. Perceived delays in answering public charges will result in attackers crying, "Cover-up!" It can be a delicate balancing act to respond urgently yet accurately. Duke seemed to accept that it would take some criticism for not responding on its attackers' timetable. In the end, it was able to provide strong evidence supporting its initial position that patients were not harmed by hydraulic fluid exposure. One must assume that it did so as quickly as possible. If Duke could have released accurate information sooner, it would have diminished some of the attacks to its image even more effectively.

CHAPTER TWO

> **Be careful pointing fingers**
> Shifting blame can be a valid strategy; however, carefully consider your scapegoat's possible response. Duke probably believed that ExxonMobil helped cause the delay in releasing to patients the chemical contents of its hydraulic fluid. In hindsight, Duke should have weighed the benefits of sharing this belief against the cost of retaliation by ExxonMobil. In the end, this choice resulted in another attack on Duke's image from ExxonMobil that was played out for the public in additional news stories.

Endnotes

1. Jean P. Fisher, "Surgical Tools Cleaned Improperly," *The News & Observer* (January 7, 2005): A1.

2. Jim Shamp, "Liquid Mix-Up at Duke Hospitals," *The Herald-Sun* (January 7, 2005): A1.

3. Sarah Avery, "Duke Slow to Find Fluid Error," *The News & Observer* (June 12, 2005): A1.

4. Jim Shamp, "Exerts: Fluid Harmful? Not Likely," *The Herald-Sun* (January 8, 2005): B1.

5. Estes Thompson, "N.C. Surgeons Unwittingly Used Dirty Tools," *Associated Press* (June 13, 2005).

6. Estes Thompson, "Surgeons at Two North Carolina Hospitals Used Instruments Washed in Hydraulic Fluid," *Associated Press* (June 14, 2005).

7. Sarah Avery, "Duke to Air Fluid Data," *The News & Observer* (June 16, 2005): A1.

8. Jim Shamp, "Duke Health Hires Lab in Fluid Mix-Up," *The Herald-Sun* (June 16, 2005): A1.

9. Valerie Bauman, "Vocal Critic of Duke Now Satisfied with Response to Fluid Mistake," *Associated Press* (June 24, 2005).

10. Victor J. Dzau, M.D., and David P. McQuaid, "Durham Regional Patient Letter 062005," DukeMedNews, *www.dukemednews.org/filebank/2005/06/102/Durham_Regional_patient_letter_062005.pdf* (accessed July 21, 2006).

11. Victor J. Dzau, M.D., and David P. McQuaid, "Durham Regional Patient Letter 062705," DukeMedNews, *www.dukemednews.org/filebank/2005/06/105/ Durham_Regional_patient_letter_62705.pdf* (accessed July 21, 2006).

12. Aaron Beard, "Duke Tell Patients Hydraulic Fluid 'Was not Harmful,' " *Associated Press* (June 28, 2005).

13. Associated Press Staff, "Exxon: Duke Didn't Say Why It Needed Hydraulic Fluid Ingredients," *Associated Press* (June 30, 2005).

14. John Stevenson, "Duke to Share Hydraulic Fluid Tests," *Associated Press* (June 28, 2005).

15. Associated Press Staff, "Duke Sued Over Surgical Instruments," *Washington Post Online*, *www.washingtonpost.com/wp-dyn/content/article/2006/07/14/AR2006071400053_pf.html* (accessed July 14, 2006).

16. Victor J. Dzau, M.D., "Update on Hydraulic Fluid Issue," DukeMedNews, *www.dukemednews.org/ filebank/2005/07/108/Dzau_memo_061505.pdf* (accessed July 13, 2006).

17. Christopher DiFrancesco, "Updated Summary of Response to Hydraulic Fluid Incident," DukeMedNews, *www.dukemednews.org/filebank/2006/01/114/Summary_of_Response.pdf* (accessed July 13, 2006).

18. Scott Wallask, "Misplaced Hydraulic Fluid Leads to CMS Citations in Three NC Hospitals," *Briefings on Hospital Safety* 13, no. 8 (August 2005): 1.

3

Children's Hospital deals with child molestation and pornography accusations

Children's Hospital in San Diego has a long history of pediatric care and advocacy, and its Chadwick Center for Children and Families has been recognized as a national authority on detecting and dealing with child abuse. However, officials at Children's were notified by law enforcement that some of its most vulnerable patients could be the victims of molestation and pornography. More shocking still, the accused was a respiratory therapist with 26 years of service at Children's Hospital. Some attacked the hospital for not doing enough to ensure the safety of the children in its care.

Chapter 3

Children's Hospital deals with child molestation and pornography accusations

About Rady Children's Hospital

Children's Hospital is a 248-bed nonprofit pediatric medical center serving San Diego and Imperial counties. According to its Web site, it has 625 physicians and 800 nurses on staff, nearly 3,000 employees, 450 active volunteers, and more than 1,200 auxiliary members. Children's 59-bed Convalescent Hospital is the only pediatric skilled nursing facility in California. With a $60 million gift, announced in June 2006, the board of trustees unanimously renamed the 52-year-old institution Rady Children's Hospital and Health Center.

Chapter three

Background of the controversy

San Diego Police Chief William Lansdowne and then Children's Hospital CEO Blair Sadler appeared together at a press conference March 9, 2006, to announce that Wayne Albert Bleyle, an employee of the hospital for 26 years, had been arrested on 40 counts of possession, manufacture, and distribution of child pornography and two counts of lewd and lascivious acts on two children under the age of 14.[1] In all, the charges involved nine patients of Children's Convalescent Hospital, but law enforcement suspected Bleyle of abusing many more. When a prosecutor asked him how many children he had abused, Bleyle reportedly said his victims equaled the number of "snowflakes."[2]

As a licensed respiratory therapist, Bleyle had unfettered access to Children's weakest patients. "Some are awake and not aware of their surroundings; others are comatose," said Pamela Dixon, director of the Convalescent Hospital. "They need hands-on care. They can't care for themselves. Many have been with us for years."[3]

Disturbing charges

Law enforcement officials said the hospital had cooperated in the investigation, which began from a tip about child pornography on the Internet. Police alleged that Bleyle used his camera phone to take pictures of bedridden children and distribute them on the Internet.[3] Once an arrest was made, Children's staff notified patients' families of the investigation and established a toll-free hotline to provide information to parents. "We took steps to immediately ensure that Bleyle would not have access to children in our care," said Sadler.

This had been the third instance of allegations of child molestation or pornography involving the medical center since 1991, and Children's officials said they conduct extensive background checks on all new employees and volunteers and provide annual training on child-abuse prevention and identification.[4] But Bleyle had no known criminal record, according to police.[4]

Moreover, Bleyle was extremely good at hiding his alleged abuse, said hospital officials. "People are shocked, dazed, and devastated," said Dixon. "They want to know: 'Was there a clue? Did we miss something?' "[2]

The state Department of Health Services and the Joint Commission on Accreditation of Healthcare Organizations began investigations of Children's Hospital. Officials said regulations about guarding against patient abuse were vague.[2] "We're continuously looking for ways to reduce the risk. But given the nature of this phenomenon and the people who engage in this behavior, [those] who think they can create an environment free of any risk have deluded themselves," said Charles Wilson, executive director of the Chadwick Center.[2]

Children's response

Nonetheless, Children's Hospital responded to the incident by announcing stricter rules, including a ban on cell phones and cameras in patient-care areas, random and frequent supervisor checks on caregivers, and a requirement for having two caregivers involved for certain medical procedures. In addition, the hospital offered families free forensic exams for patients.

Although Children's received support from some in the community, others attacked the organization: "Here they are, this great protector of kids, and they can't protect

Chapter three

their own," said Milton Silverman, an attorney specializing in child abuse cases. "That they could come up with any kind of excuse as to why this happened is to me absurd and unconscionable."[2]

While Children's Hospital and the community were still reeling from the accusations against Bleyle, the police alerted the hospital that another employee was under investigation. Christopher Alan Irvin was arrested on April 14, 2006, and charged with twice molesting a semicomatose four-year-old girl. According to police, Irvin, a nurse who worked at Children's for a year and a half, admitted that he abused the dying child behind closed curtains in the intermediate care unit where he worked.[5]

Law enforcement officials said the two incidents at Children's were unrelated and that there was no evidence that Bleyle and Irvin knew each other.

In a prepared statement to the press, Children's Sadler reacted with anger: "Words cannot describe how furious we are that this could happen at Children's Hospital. All of us—our patients, our families, and our staff—are still in the healing process from the Bleyle case. We are outraged at this second betrayal of trust by someone in our midst."

The hospital's statement went on to reiterate Children's thorough process for screening new employees and pointed out that both Bleyle and Irvin had "completely clean records, which is not uncommon for this type of perpetrator."[6] Even though Children's suggested that it took reasonable measures to protect patients, it promised to do more to enhance security.

Additional safety efforts

Children's Hospital appeared to take its efforts toward improving surveillance seriously. In May 2006, it announced that it had suspended three physicians because an audit of the hospital's computer system detected that someone had used an access code to visit a pornography site. Taking no chances, hospital officials turned the information over law enforcement's Internet Crimes Against Children Task Force. "We're just so sensitive to these things now that we immediately turned over the information," said Children's spokesman Tom Hanscom.[7]

Sadler announced his plans to step down as CEO of Children's to pursue new opportunities working on quality improvement and health policy issues involving children at UCSD School of Medicine and Management and the Institute for Healthcare Improvement. A *Union-Tribune* article highlighted the many challenges facing newly announced CEO Kathleen Ann Sellick. The article said she faced not only the difficulties of the two molestation and pornography incidents but also a problematic financial picture that includes dealing with "a two-year standoff over a labor contract for 700 hospital members of the Service Employees International Union, Local 2028."[8] Seizing an opportunity in the press, one union member suggested the hospital's recent layoff of staff made it more vulnerable to incidents of child abuse. "We are running very short-staffed. There aren't enough people walking in and out of the rooms," said Jane Stellenwerf, a medical surgery floor secretary. "I'm hoping [Sellick] will listen to our plight."[8]

Throughout these ordeals, Children's Hospital officials expressed shock and outrage. They claimed that despite their best efforts at implementing systems to protect patients from sexual predators, there is little that any organization could do to prevent such incidents. Officials pointed out that similar situations have occurred at other institutions such as schools, churches, and community centers.[2]

CHAPTER THREE

If Children's officials worried that these incidents might negatively affect philanthropy to the hospital, these concerns were soon eased. The hospital announced in June 2006 that the Rady family had given Children's the largest donation in its history. In recognition of the $60 million gift, the board of trustees voted in favor of renaming the medical center the Rady Children's Hospital and Health Center.

Children's stakeholders

In dealing with two alleged child abuse crimes, the following groups influenced Children's public image:

- Hospital leaders
- Law enforcement
- Children's PR department
- Staff
- Donors
- Accused child molesters (former employees)
- Alleged victims and their families

Throughout this sensitive situation, Children's interacted with the above stakeholders. Upon learning of charges against Bleyle and Irvin, Children's terminated their employment. It also reached out to the families of alleged and potential victims and created a dedicated hotline to receive inquiries from parents. In addition, Children's communicated with physicians and staff about the incidents and about the policy changes it was implementing in response.

Analysis of efforts to restore Children's public image

Delicate maneuvering is urged for the organization that attempts to respond to charges that an employee has committed abuse on a patient. Children's Hospital had to assist law enforcement in its investigations, interact with concerned family members, address employee morale, and reassure the community that it takes patient safety seriously. In these efforts, Children's attempted to defend its public image with the following image restoration strategies: denial by shifting the blame, evading responsibility by showing good intentions, corrective action, and reducing the offensiveness of the act by minimization and bolstering.

Shifting the blame

Children's Hospital had obvious scapegoats, and its officials smartly attempted to shift the blame on the individuals accused of the crimes. According to law enforcement, Bleyle and Irvin confessed to the crimes, as reported by the *Union-Tribune* (the pair later pleaded not guilty to charges). Nevertheless, Children's made efforts to drive the point home. "Sexual predators are found in all aspects of life, often masquerading as caring and responsible individuals while seeking covert opportunities to gain access to victims," said Wilson, who Children's identified as a nationally recognized expert in child maltreatment. "Despite the best efforts of organizations and communities, we can never completely eliminate the risk of this type of illegal activity."[4]

In an announcement of Bleyle's arrest, Dr. Cynthia Kuelbs, chief of the medical staff at Children's, further attempted to point out that this offensive act was committed by an individual: "We are outraged that an individual who has walked among us, helping heal children day in and day out, would be capable of betraying those same children, their families, coworkers, the mission of the hospital and the entire community."4

The hospital's director of pastoral care, Rev. John Breding, even suggested that Bleyle was looked to as a leader by some. "People with less experience would go to him," he said.9 This suggests that Bleyle had succeeded in deceiving coworkers about his motives.

In a letter published in the hospital's quarterly magazine, Sadler, the outgoing CEO, suggested that Bleyle victimized his colleagues as well as his patients: "As one would expect, the families involved have reacted with shock, grief, and anger—feelings that all of us here and throughout the community share. All our efforts now are concentrated on supporting our families and our caregivers and helping them to heal from this horrible betrayal."10

Attackers who said Children's could have prevented these incidents were few. Many others made comments to the media supporting the hospital's claim that even an institution that puts forth reasonable efforts to protect its vulnerable patients can suffer from the actions of disturbed employees. Children's efforts to shift the blame to Bleyle and Irvin were effective in this regard.

Good intentions

Through words and actions, Children's officials displayed their good intentions toward patients, their families, and the community. Perhaps most prominently, Children's CEO immediately made his presence felt in the initial news conference with police. The image of a hospital's CEO standing next to the chief of police can be quite powerful. It displays that the organization is partnering with law enforcement, sharing information with stakeholders, and addressing the problem head on.

"Our families and staff share a common bond to protect and nurture these vulnerable children. The prospect of this type of crime in our midst is contrary to everything we believe in and strive to achieve every day," said Children's Dixon.[4]

In addition, Children's officials made an effort to show how they cooperated with law enforcement and provided information to salient stakeholders. "The transparency and openness with which the administration and the board have addressed this issue are exemplary," said incoming CEO Sellick. "Continued open communication will be essential, and finding even more ways that we can set the standard as the safest environment for children."[8]

It is unreasonable to suggest that an institution like Children's would have anything but the best of intentions for its patients. If its patients are placed at risk, its own reputation is equally at stake. Therefore, attempts to promote its good intentions appear to have been well received. They have been included in media reports and reiterated by outside sources.

Corrective action

Rhetorically, it can be difficult to deny responsibility for an offensive act and then announce corrective action. After all, if one is not responsible for an offense, then what is there to correct? However, Children's framed its corrective action strategy well—even though it insisted that Bleyle and Irvin were responsible for the offensive acts. First, the hospital removed the accused individuals from patient care. Next, it launched an internal investigation to uncover opportunities to improve patient safety, and it set up a hotline for concerned parents.[4]

After the second incident of alleged child abuse, Children's announced plans for a host of actions to enhance security, including the following:[6]

Chapter three

- Chaperones will be assigned to each patient unit to ensure that personal aspects of care are accompanied by two staff members when a curtain or door needs to be closed for privacy or when parents are not present

- New technologies will be used to enhance ongoing background checks of all employees

- Curtains and doors for all patients will be left open, except as specifically requested by patients or families

- Clinicians will make random unannounced rounds in all patient areas

- No photography of patients by staff will be permitted, except as authorized by hospital policy

- Mirrors will be installed in areas not readily visible on rounds

- Higher visibility patient curtains will be installed in all appropriate areas, shielding caregiver staff only from chest to ankle

- Staff will not be allowed to bring personal cell phones into clinical areas

- Real-time audits of Internet use will be conducted on all Children's computers to detect access to pornographic sites

These efforts may have helped Children's spot an employee's noncompliance with the hospital's Internet policy. The hospital suspended three physicians when an audit of the computer system discovered that someone had used an access code to visit a pornography site. Children's officials also alerted law enforcement of the incident. A story by the Associated Press suggested that Children's has become even more vigilant in its patient protection efforts.

Minimization

Children's Hospital attempted to minimize its responsibility for the offense by stating that it had made reasonable efforts to protect its patients. Officials said that when they consider an applicant, they screen Social Security numbers, check for aliases, review professional references, and examine criminal records and sex-offender databases.[2]

Children's noted to the press that both Bleyle and Irvin had clean criminal records, and both were licensed for their professions. A prepared announcement about Irvin's arrest also suggested that the organization was not responsible for these acts by individual employees because of the safety measures it already had in place. "Even before the Bleyle case, we had established very high standards in screening procedures for hiring new employees. However, we believe that even the most stringent background checks alone are not enough. Both Bleyle and Irvin had completely clean records, which is not uncommon for this type of perpetrator."[6]

Children's position received support from a lawyer with the California Healthcare Association. "The logistics of obtaining every patient's consent for 24/7 surveillance clearly make this idea impossible," said Lois Richardson. "Although what happened at Children's Hospital is reprehensible, the truth is these situations don't happen very often."[2]

When reports of patient abuse become public, the knee-jerk reaction is to question whether the institution is to blame. In Children's case, state regulators and the

Joint Commission immediately announced that they would investigate the hospital. Children's adequately addressed some of their concerns by showing how it follows industry standards with regards to background checks. Coupled with its strong efforts of shifting the blame to the accused, Children's implemented an effective minimization strategy.

Bolstering

Communicating with stakeholders during a time of shock, anger, and pain is extremely difficult. The organization must respond to the public's concerns and work to enhance the safety of those with whom it has been entrusted. This leaves little room for the organization to attempt to remind the public of its many good deeds. Moreover, this sort of effort might not be well received during a time of increased scrutiny. When done correctly, however, a bolstering strategy can express to salient stakeholders the organization's desire to continue its mission of serving the community. Children's expressed this sentiment well when opportunities presented themselves.

For instance, after the second arrest for alleged molestation of a Children's patient, the hospital communicated many enhancements to its safety policies. It went on to remind the press that it takes its commitment to the community seriously. "Our mission—to restore, sustain, and enhance the health and development potential of children through care, education, research, and advocacy—these are not just words on a piece of paper, or on a wall. We live these words every day. They guide everything we do. These two incidents are profound contradictions to our mission, especially for the heroic caregivers who practice our mission every day. Children's wishes to assure this community that your doctor, your nurse, your caregiver are here to heal your child."[6]

Conclusion

The prevalence of the media's reporting of crisis events has one intangible benefit for healthcare organizations: It has opened the public's eyes to the fact that immoral people may do horrible things at good institutions. Even though an organization should recognize that an employee can have the potential for exposing it to these sorts of scandals, Big Brother–type measures to reduce risks can create additional problems. For example, placing a surveillance camera on each patient could open the organization to patient privacy concerns. Children's Hospital did many things correctly in its implementation of image restoration strategies, and its recent $60 million gift can be considered a positive sign that the damage to its public image from these incidents are unlikely to be long lasting.

Implications for action

This case study presents a scenario that has the potential for occurring in just about any organization. Considering that healthcare institutions are places where people go to mend, revelations of patient abuse can have an especially chilling effect on a community. Here are a few tips from the Children's case:

Cooperate with law enforcement

Not only is it wise to comply with a law enforcement investigation for legal reasons, it sends your stakeholders a strong message that your organization is on the side of the law. When someone from your organization is under criminal investigation, you want your senior leaders to appear as partners with law enforcement. This can help transfer the good will generated by law enforcement to your leaders. Children's Hospital accomplished this by ensuring that its CEO was present and made a statement with the chief of police at the initial press conference. Photos of the CEO and police chief standing side by side are far preferable to an organization than photos of a current or former employee being dragged off in handcuffs.

Display transparency

In situations of alleged patient abuse, the public will have many questions, and a healthcare organization should recognize the public's right to truthful and complete answers. Responses to stakeholders and the media should be carefully crafted but direct and accurate. Review the literature in the press and on Children's Web site for comments made by senior leaders at Children's. Their statements are sincere, are thoughtful, and help create a strong counter-narrative defending the organization.

Differentiate the actors from the organization

This was a case in which the organization had to shift the blame on the alleged perpetrators. Although this may seem like a commonsense statement, keep the public focused on those accused of the crime and point out that criminals have been found in all walks of life. Children's Hospital clearly succeeded in these efforts.

Fill the gaps

The public may forgive an organization for dastardly acts committed by individual employees; however, the hospital must realize that this brings a new level of scrutiny on the organization. It cannot afford repeat offenses to occur. The public expects more than just words in response to these incidents; it expects the organization to redouble its efforts to protect those in its care. Children's seems to have made strong efforts to enhance its safety measures. Time will tell whether these are sufficient. If another crime of this scope is alleged, Children's will find itself under intense public pressure.

Show your good side

When senior leaders speak about discoveries of patient abuse, they should make a point of reassuring the public that the incidents will not distract the organization from continuing its mission of serving the community. These statements might get filtered out by the media, so make an effort to take them directly to stakeholders—including patients, families, donors, board members, and employees.

A predictable attack

If you are dealing with a crisis situation at the same time that you are negotiating with a union, expect that a union representative will exploit this opportunity to damage your reputation. In Children's case, a union member suggested that the hospital's layoff of workers likely contributed to incidents of alleged child molestation. The organization is well advised to not single out these attacks. Instead, stay on message with your image restoration strategies.

Endnotes

1. Penni Crabtree, "Therapist at Children's Hospital Accused of Child Molestation, Porn: Hospital's Quick Response Crucial, Crisis Experts Say," *The San Diego Union-Tribune* (March 10, 2006): A1.

2. Cheryl Clark, "Questions Pile Up in Alleged Abuse at Children's Hospital," *The San Diego Union-Tribune* (March 15, 2006): A1.

3. Joe Hughes, "Therapists at Children's Hospital Accused of Child Molestation, Porn: Longtime Worker Arrested; Charges Involve 9 Youths," *The San Diego Union-Tribune* (March 10, 2006): A1.

4. Tom Hanscom, "Children's Convalescent Hospital Employee Arrested," Rady Children's Hospital, *www.chsd.org/body.cfm?xyzpdqabc=0&id=43&action=detail&ref=74* (accessed July 17, 2006).

5. Ray Huard, "Ex-Nurse Admits Abusing Dying Girl, Warrants Show," *The San Diego Union-Tribune* (June 14, 2006): B1.

6. Tom Hanscom, "Children's Hospital Employee Arrested," Rady Children's Hospital, *www.chsd.org/body.cfm?xyzpdqabc=0&id=43&action=detail&ref=76* (accessed July, 17, 2006).

7. Associated Press, "Doctors Suspended Among Porn Probe," *Associated Press* (May 13, 2006).

8. Keith Darc, "Challenges Await New Hospital CEO," *The San Diego Union-Tribune* (April 23, 2006): H1.

9. Linda Orlando, "Hospital Worker Charged with Molesting Comatose Children," Buzzle.com, *www.buzzle.com/editorials/3-17-2006-91324.asp* (accessed July 17, 2006).

10. Blair L. Sadler, "Moving Forward, Moving On," *Children's* (Spring 2006): 3.

Kaiser Permanente's Northern California kidney transplant program falls under attack

With its embattled kidney transplant program in Northern California, Kaiser Permanente implemented multiple strategies to restore its image. Numerous critics charged Kaiser—whose marketing motto is "Live Well and Thrive"—with not only incompetently administering the program, but also with making unethical medical decisions. In sum, critics created stories in which Kaiser was responsible for prolonging the suffering of patients on its kidney transplant waiting list, which might have resulted in the deaths of some patients who otherwise would have received kidneys at other transplant programs.

CHAPTER 4

Kaiser Permanente's Northern California kidney transplant program falls under attack

About Kaiser Permanente

Kaiser Permanente is the largest health maintenance organization in the country, with 8.4 million enrolled members. Headquartered in Oakland, CA, it encompasses the not-for-profit Kaiser Foundation Health Plan, Inc., Kaiser Foundation hospitals and their subsidiaries, and the not-for-profit Permanente Medical Group. It operates in nine states and the District of Columbia. According to a 2004 financial statement posted on its Web site, Kaiser Permanente's total net worth is $22.4 billion. It has 134,000 employees, 11,000 physicians, 30 medical centers, and 431 medical offices.

CHAPTER FOUR

Background of the controversy

At its most basic level, it made perfect business sense for Kaiser to operate its own kidney transplant program from the Kaiser Foundation Hospital in San Francisco. With the advent of less invasive procedures, the hospital's open-heart surgery program had unused capacity.[1] Before opening its kidney transplant program, Kaiser Permanente of Northern California members received kidney transplants at well-established University of California centers, UC Davis, in Sacramento, and UC San Francisco hospitals. Kaiser had contracts with these hospitals and paid them approximately $65,000 per kidney transplant.[2] In mid-2004, Kaiser notified approximately 1,500 members that their kidney transplant care would be transferred to the Kaiser Foundation Hospital. According to the *Los Angeles Times*, this kidney waiting list was one of the longest in the country, and no other kidney transplant program had established a waiting list of this size so quickly.[1]

After less than two years of operation, Kaiser's program became bogged down in a scandal when the *Los Angeles Times* and television news CBS 5 (KPIX-TV) in San Francisco reported that for 2005, the program's first full year of operation, Kaiser performed 56 transplants and twice as many people had died on its waiting list; at other transplant centers in the state, more than twice as many patients received kidneys than died.[1] *Times* reporters Charles Ornstein and Tracy Weber pointedly criticize Kaiser's program: "Hundreds of patients were stuck in transplant limbo for months because Kaiser failed to properly handle paperwork. Meanwhile, doctors attempting to build a record of success shied away from riskier organs and patients, slowing the rate of transplants performed."[1]

Trust lost

In this and subsequent articles, the newspaper told heart-wrenching stories of distressed Kaiser patients stuck between two kidney transplant programs while inept and understaffed administrators bungled their cases. The stories painted a picture of patients hooked up to dialysis machines slipping into despair and family members left wondering whether Kaiser's mismanagement led to the loss of loved ones. Moreover, the *Los Angeles Times* suggested that Kaiser did not communicate problems to patients. "I have no confidence in [Kaiser's] transplant people," said Kaiser patient Jason Mitchell to Weber and Ornstein. "I don't want them touching me."[3]

These accusations are backed up by statements to CBS 5 by the program's former administrator, David Merlin. "Many, including myself, were concerned that some patients may have had a tremendous decline in their health status because of the neglect that had occurred," he said. "And the patients have a right to know."[4]

Perhaps more disturbing, Ornstein and Weber reported that Kaiser denied "nearly perfectly matched kidneys" offered to 25 of its patients by UC San Francisco.[5] With comments from a medical director at UC San Francisco, a Kaiser transplant physician on leave, and a bioethicists, the article suggested that Dr. Sharon Inokuchi, medical director for Kaiser's program, did not make the ethically wise medical decision in rejecting the 25 kidneys without informing patients.

In the following weeks, more critics emerged, state and federal investigations began, and lawsuits against Kaiser were filed. The day after the director of California's Department of Managed Health Care told the *Times* that Kaiser would pay for transplants at University of California hospitals and was quoted saying Kaiser would "do what the patients want them to do,"[2] Mary Ann Thode, regional president of

CHAPTER FOUR

Kaiser Foundation Health Plan and Hospitals, publicly apologized to patients who felt that Kaiser had ignored their complaints.[6]

Closing time

The *Los Angeles Times* editorial staff made Kaiser Permanente the poster child for transplant programs in need of stricter regulatory oversight industrywide.[7] Within 10 days of the first *Times* story, Kaiser announced its plans to suspend its Northern California kidney transplant program indefinitely and said that it would transfer the patients on its transplant waiting list, which had grown to approximately 2,000 patients, to the University of California hospitals.[8] However, after this announcement, even more rhetorical attacks were made on Kaiser's image.

In other press reports, regulators, patients, and a UC medical leader were critical of Kaiser's communication with patients. As early as May 6, 2006, Kaiser officials stated that they planned to contact the patients on its kidney transplant waiting list in an effort to address their concerns.[3] Upon Kaiser's announcement of the closure of the transplant program, the state's director of the Department of Managed Health Care said that Kaiser was required to maintain a 24-hour toll-free hotline to respond promptly to patients.[8] In addition, the medical director of renal transplant services at UC San Francisco, which would receive most of the patients from Kaiser, said that Kaiser officials told him the HMO would phone all of its transplant patients to update them about their cases. "We were led to believe that Kaiser was going to be actively calling patients to explain to them what was going on with the transplant program," said Dr. Stephen Tomlanovich of UC San Francisco.[9] However, a Kaiser spokesman said on May 17, 2006, that the organization had no plans to call patients and that patients would soon receive a letter explaining the transfer process. In the same *Times* article, reporters Ornstein and Weber wrote that "days after Kaiser and regulators promised patients a smooth transition to outside programs, the *Times* found that anxious callers had trouble getting through to a 24-hour hotline or had been given baffling, incorrect, or useless information about their care."[9]

Accusations of mismanagement

Considering Kaiser had already announced that it was suspending its kidney transplant program, perhaps the practical effects of a federal government report critical of Kaiser were limited. If nothing else, the report from the U.S. Centers for Medicare & Medicaid Services allowed reporters to repeat all of the accusations and criticisms that could have a substantial effect on Kaiser's image. In short, the report accused Kaiser of inadequately planning, staffing, managing, and assessing its Northern California kidney transplant program. Regulators said that Kaiser remained out of compliance with government standards and could lose federal funding for end-stage renal patients.[10]

Kaiser faced further scrutiny about the amount of time it took to transfer its kidney transplant patients to UC hospitals. Differences in recordkeeping complicated the process, according a deputy director of the California Department of Managed Health Care.[11] However, the probable lasting impression from this story was that Kaiser had promised to swiftly and efficiently transfer the more than 2,000 patients, but at the time of the *Times* report—more than a month later—only one Kaiser patient had received a kidney at UC Davis.

Altogether critics' attacks from May 3 to August 11, 2006, potentially conspired to create lasting damage to Kaiser Permanente's reputation. Even at the time of this writing, Kaiser has not yet transferred all of the kidney transplant patients on its waiting list. The California Department of Managed Health Care has broadened its investigation to determine whether Kaiser's Northern California operation routinely ignored or mishandled patient complaints. The HMO regulator has also fined Kaiser Permanente $2 million; in addition Kaiser has agreed to donate $3 million to support a nonprofit organ and tissue donor registry program. Putting the quantitative impacts of this crisis event aside, a qualitative review of the literature in the press—print, broadcast, and online—suggests that many accusations directed toward Kaiser's administration of its Northern California kidney transplant center have merit and have been substantiated by a government review; therefore, the long-term effect that this scandal could have on the larger organization warrants a serious, well-directed restorative strategy.

Kaiser's stakeholders

Pubic comments and related actions by those mentioned below have influenced the public's opinion of Kaiser:
- Senior leaders
- Administrators of the kidney transplant program
- Medical staff
- Company spokespersons
- Patients and their families
- State regulators
- Federal regulators
- Media

Kaiser Permanente's Northern California kidney transplant program falls under attack

> There are a number of communication channels at Kaiser's disposal as it seeks to influence salient stakeholders. Spokespersons, senior leadership, and clinical staff have interacted with media members. In addition, Kaiser has communicated about the issue internally to its employees, according to Matthew Schiffgen, a spokesman for the HMO. He said that Kaiser provides information to state regulators on a weekly and as needed basis and that it has sent its kidney transplant patients letters informing them about the transition process back to University of California hospitals. The letters also notify readers about Kaiser's toll-free hotline. On June 15, 2006, it published a Web site to provide patients and their families with regular updates about their care.
>
> The news media raised legitimate concerns about the kidney transplant program, said Schiffgen. However, he questioned whether members of the press—given their space and time constraints—have provided readers with a full understanding of this complex situation. Even though Kaiser's communication professionals continue to collaborate with the media, he said that they also use alternative methods to communicate with key stakeholders.

Analysis of efforts to restore Kaiser's public image

As Benoit points out, an attack on one's image must include two components: An act has to be considered offensive by salient stakeholders, and the individual or organization accused has to be held responsible.[12] Clearly, Kaiser's Northern California kidney transplant scandal fits precisely within this definition. All stakeholders listed in the section above acknowledged problems with the program and Kaiser's responsibility for the offensive act.

CHAPTER FOUR

Kaiser officials attempted to defend the organization's image by using all five broad categories of image restoration strategies. They often used multiple strategies in a single statement. Below I show how Kaiser Permanente, in an effort to influence stakeholders' perceptions of the company, executed these image restoration strategies: denial, reducing offensiveness of the event by minimizing and bolstering, corrective action, mortification, and evading responsibility by showing good intentions.

Denial

The initial *Los Angeles Times* report effectively linked administrative and medical decisions made by Kaiser representatives to bad outcomes for kidney transplant patients. It showed how errors in administration greatly extended the duration that patients spent waiting for kidneys. It also implied that some Kaiser patients may have died while waiting for kidney transplants and that these patients might have received transplants in time at the University of California hospitals had their care not been disrupted.[1]

In response to this attack, the *Times* reported that Kaiser officials denied there were problems. And the program's head transplant surgeon, Dr. Arturo Martinez said, "Everything has been going on track."[1] This quick denial in the face of the newspaper's evidence was not effective. In the same story, Kaiser officials admitted that information it provided to the *Times* was incomplete or misleading. Moreover, Bruce Blumberg, Kaiser's physician-in-chief, conceded that serious problems with the program existed.[1]

Kaiser denied the validity of other attacks on its image as well. When UC San Francisco's Tomlanovich was concerned that the records for more than 200 patients were not properly transferred to Kaiser, he claimed that his hospital contacted Kaiser officials by phone, fax, and e-mail. However, Inokuchi, the medical director for Kaiser's transplant program, denied receiving these communications.[1] Tomlanovich also accused Kaiser of turning down 25 cadaver kidneys that were "perfectly matched" to patients on Kaiser's waiting list. According to the *Times*, an e-mail from Tomlanovich to a UC San Francisco colleague in February 2006 corroborated his claim.[5] Kaiser officials initially denied Tomlanovich's accusation. Weber and Ornstein wrote that "Kaiser officials adamantly denied that they had ever instructed UC San Francisco to turn away such organs. But after being confronted with evidence to the contrary by the *Times*, the officials called back to say that they could not stand by that position."[5] Tomlanovich's claims were substantiated further by accounts from a Kaiser kidney specialist, Dr. W. James Chon, who said that Inokuchi would not give permission when he inquired about one of these transplants.[5]

The *Times* story then offered this stinging assessment of Kaiser by healthcare expert Arthur Caplan, director of the University of Pennsylvania Center for Bioethics: "Something is simply out of whack with the health plan's priorities."[5]

Perhaps most damaging, some patients on Kaiser's kidney transplant waiting list and family members of patients shared their beliefs that Kaiser's mismanagement resulted in deaths. Ruben Porras died while on Kaiser's waiting list. Before his case was transferred, he had spent three years on the waiting list at UC Davis.[1] "It is likely that he would have been transplanted fairly soon," said Dr. Richard Perez, chief of the UC Davis transplant center.[1]

The *Times* was very careful to note on several reports that, at the time, it was not possible to prove patients died as the result of Kaiser's errors.[5] However, media sources said that deaths likely occurred because of these errors. For example, commenting on a public apology by Kaiser's top official in Northern California, patient

Chapter Four

Bernard Burks said, "The guys who died [awaiting their transplants], they don't hear the apology at all."[6]

Thode, the top official, offered a thoughtful and measured response to questions about whether foul-ups caused patient harm. "I certainly hope that no one was harmed or died," she said. "I want to finish the initial review before I make any comment."[6] Other Kaiser officials denied that patients were harmed at all. Inokuchi and Martinez told federal investigators that the transfer to Kaiser's kidney transplant program did not impair patients' opportunities for transplants; however, the federal report about Kaiser's program contradicted their claims.[10]

It is inconclusive whether Kaiser's errors will be directly linked to patient deaths. Simple denials from Kaiser officials do not appear persuasive, considering previous misstatements made to the press and denials that were contradicted by federal investigators.

Minimization

The strength of the media reports about Kaiser's scandal were based largely on the commonsense premise that patients were likely harmed because of administrative and clinical mismanagement. That is, if in the court of public opinion most reasonable people believe that these problems did not harm people who entrusted Kaiser with their care, this story would not appear multiple times on the front pages of newspapers and on TV news programs. In fact, if Kaiser successfully refuted the claim that patients were harmed, the likely outcome would have been that the scandal would not have developed at all.

Therefore, it is not surprising to see organizations gripped with scandal attempt minimization strategies. Kaiser's Blumberg claimed that care for patients had not been affected by problems at the transplant program.[1] In an interview with CBS 5, Inokuchi went further. She said that administrative problems resulted in lost patient records, but this could have been prevented by the patients themselves. "There were very few [patients] who complained," she said. "I think the patients who were lost were patients who were late to join Kaiser or did not respond to communications from their own physicians."[13] This comment could also fit in the image restoration strategy of shifting the blame, in this case to the patients who claimed that they were victimized by Kaiser.

Statements from Kaiser officials that patients were unharmed because of errors in the management of its kidney transplant program were ineffective and ran the risk of public backlash. Several anecdotes from patients created a narrative in which they have in fact suffered physically from prolonged dialysis treatment and were under emotional distress in the wake of the scandal. Although it is natural for members of an organization under attack to take a defensive posture, doing so can exacerbate the damage to the organization's image. An exceptional example of this is a statement by a physician with the Kaiser-affiliated Permanente medical group in Northern California. "It isn't as if waiting another six months or nine months for a transplant is a death sentence," said Dr. Sharon Levine, associate executive director for Permanente. "There are patients who choose to stay on dialysis rather than going through a transplant."[5] In a follow-up story from the *Times*, a Kaiser patient undergoing dialysis sessions three days a week responded angrily to Levine's comments.[3]

It did not take long for Kaiser's attempts at minimization to haunt it in another way. Its minimization attempts were quickly discredited when state and federal regulators harshly scrutinized the HMO. An agreement with the California Department of Managed Health Care required Kaiser to pay for its dissatisfied patients' kidney transplants at other hospitals.[2] Within three days of this news,

Kaiser announced that it was indefinitely suspending its kidney transplant program.[8] It defies reason that Kaiser would provid high-quality patient care that did not lead to negative outcomes for its kidney transplant patients and then voluntarily ended this program in the midst of controversy.

Often when organizations deal with crisis, senior leaders will attempt to distance themselves from the scandal. This might be done to ensure the continued effectiveness of the leader within the organization and to distance the senior leader from future legal fallout. When Thode, regional president of Kaiser Foundation Health Plan and Hospitals, offered a public apology to the kidney transplant patients, she also minimized her direct involvement in the scandal. "I am not personally aware of complaints that have come in regarding the kidney transplant program. I have actually received numerous letters from patients who have actually been happy with the care."[6]

The press conference during which she made this statement occurred more than a week after the news of mismanagement at the transplant center broke on CBS 5 and in the *Los Angeles Times*. A reasonable person would, therefore, expect the senior leader of Kaiser at this press conference to be aware of patient complaints by this time. Other statements from Thode appear well measured and appropriate, but this comment suggests that in Kaiser's corporate culture, doctors and staff do not escalate patient complaints through the hierarchy of the organization in the same way they do letters of gratitude.

Bolstering

It is natural to think that bolstering would be a particularly strong image restoration strategy for a healthcare organization under attack. After all, healthcare institutions protect our most cherished asset—our lives—so bolstering could be an effective way to generate good will among stakeholders. Kaiser shared an important fact with the media about its kidney transplant program. Although it grudgingly acknowledged that it performed fewer kidney transplants on the patient population previously treated at the UC hospitals, it boasted the fact that it had not recorded a patient death after transplant surgery since the program had opened.[1]

However, there were two problems with this attempt at bolstering. First, Kaiser accepted fewer donated kidneys on behalf of its patients than did its neighboring programs.[1] The reader is left to infer that Kaiser physicians intentionally shied away from risky donors to boost the program's survival rate, even though many Kaiser patients had signed off on accepting riskier kidneys when they were in the UC programs.[1] The other problem with the exceptional survival rate that Kaiser officials bragged about was that it was not true. In a later article, Kaiser told reporters that it erred when it said that none of its patients died since the transplant center opened; in fact, one patient had died after kidney transplantation.[14]

"I am delighted to be joining such a forward-thinking organization and feel honored to partner with this strong team to further enhance the Kaiser Permanente brand," said Diane Gage Lofgren, the newly appointed vice president of brand strategy, communications, and public relations for Kaiser Foundation Health Plan, Inc., and Kaiser Foundation Hospitals, according to a prepared statement.[15] Perhaps after reading some 3,000 words about Kaiser's kidney transplant program scandal, this quote might seem insincere, but it is important to remember that in many other aspects Kaiser Permanente is considered an innovator in the healthcare industry. In this context, Lofgren's comments are not at all inappropriate. For example, the Kaiser Foundation Health Plan of Northern California was ranked 58th out of more than 500 health plans in the *U.S. News & World Report's Best Health Plans of 2005*.[16]

Lofgren's press release did not include any mention of the kidney transplant program scandal, nor would this be recommended. However, her bolstering strategy went unnoticed when the *San Francisco Business Times* ran a short article about her appointment. The paper mentioned Kaiser's troubles: It said that Lofgren would oversee "Kaiser's highly successful Thrive advertising campaign, while working to rebuild the system's tarnished image after a series of recent blows, many of them connected to serious operational and management problems at its Northern California kidney transplant unit, which is in the process of being closed."[17]

Kaiser's bolstering strategies were ineffective. When it boasted about the program's patient-survival rate, it invited more criticism; when it had to back off of its claims, it gave critics another chance to point out that Kaiser officials provided the press with inaccurate information. Even a rather harmless press release did not effectively incorporate this strategy and resulted in yet another press account of the scandal.

Corrective action

Among the first accusations that Kaiser acknowledged was that it had staffing problems at the transplant center. Ten employees of a staff of 22 resigned or were terminated since the program opened. In addition, Inokuchi was left as the only kidney specialist (or nephrologist) at the hospital; experts in the field said the program should have had at least four or five.[1]

Blumberg said that Kaiser was planning to improve upon the number of surgeries to about 90 transplants in 2006.[1] His optimistic comment was echoed by Levine, who said the program was building capacity.[18] Of course, these plans would not be realized because Kaiser decided to suspend the program and transfer patients to UC hospitals. Cindy Ehnes, director of the California Department of Managed Health

Kaiser Permanente's Northern California kidney transplant program falls under attack

Care, offered this biting analogy about Kaiser's decision: "They were trying to essentially fix the airplane while they were flying it."[8]

In other attempts to implement its corrective action strategy, Kaiser sought to display how it was responding to the scandal by communicating with the patients on its kidney waiting list. On May 6, 2006, the *Times* reported that Kaiser officials planned to contact the estimated 2,000 patients on its waiting list to tell them how to seek information from the program.[3] Kaiser made available a toll-free hotline for these patients.

These attempts at corrective action could have had a positive effect on Kaiser's image insomuch as they showed the organization responding appropriately to concerned patients, but these efforts resulted in further scrutiny from stakeholders. Kaiser sent patients a letter dated May 22, 2006[19]—20 days after the news of administrative and clinical mismanagement broke—informing them about its plans to transfer them to the UC hospitals. Throughout these 20 days, some anxious Kaiser patients saw the scandal unfold through dramatic stories in the newspapers and on TV news without receiving information about their care directly from Kaiser. Furthermore, California regulators ordered Kaiser to correct problems with the hotline it created after learning that phone staff could not answer patients' basic questions.[9]

In notifying patients and in statements to the press about the plans to transfer transplant patients back to UC hospitals, Kaiser officials said patients would be transferred as quickly and smoothly as possible.[19] Although these statements represent initial steps toward ending the program, they also show Kaiser officials trying to take corrective action by ending the perception that Kaiser patients awaiting kidney transplants continued to receive substandard care. This is perhaps Kaiser's strongest effort to restore its image. If it closes the program that invited scrutiny from salient stakeholders, it begins the conclusion of the scandal. In addition, it recovers some moral authority by putting the care of patients ahead of perceived organizational goals. A month after Kaiser's Thode announced these plans, the *Times* criticized the organization's transfer efforts, writing that Kaiser had transferred only "a handful" of the patients on its waiting list and that only one of these patients had transplant surgery.[11] State HMO regulators provided some support to Kaiser officials by saying that the transfer process was complex and required the three hospitals to standardize patients' medical records.[11]

Mortification

If salient stakeholders consider an apology sincere, they may choose to forgive the offensive act. In Kaiser's case, a public apology issued during a May 12, 2006, press conference came after reports of Kaiser officials providing inaccurate statements to the press about the quality and effectiveness of its kidney transplant program. Within 10 days of mostly negative news coverage about the scandal preceding it, the brief apology's effectiveness was most likely dulled. However, Kaiser made a wise move by selecting Thode to deliver the apology. As Kaiser's top Northern California official, her presence communicated that the organization was taking the matter seriously. Thode was also not directly linked to the scandal in previous news accounts. "I deeply apologies to anyone who feels that they made a complaint and that that complaint was not noted or handled appropriately," she said.[13]

Good intentions

It would be irrational to believe that Kaiser Permanente purposely provided poor administrative and clinical care to its kidney transplant patients. In all likelihood, Kaiser officials thought that the new program would serve its patients adequately

Kaiser Permanente's Northern California kidney transplant program falls under attack

while keeping expenses in check for kidney transplants. Overall, Kaiser's messages about the company's good intentions became more of a centerpiece of its communication efforts as the scandal dragged on. But these messages competed against many other statements in the press that attacked Kaiser's reputation. Much like Thode's public apology was impaired by these attacks, so too was Kaiser's efforts to reassure stakeholders that it had good intentions with regard to starting up and administering the transplant program. After charges of lost medical records that likely caused significant delays to transplant procedures, after attacks from worried patients who said their concerns were ignored, and after Chon questioned the ethics of Inokuchi's clinical decision-making, patients on Kaiser's kidney transplant waiting list received the May 22, 2006, letter that began as follows: "At Kaiser Permanente, your health is our first concern. We are writing to keep you informed about some recent changes in our Kaiser Permanente San Francisco kidney transplant program, to begin to answer your questions, and to describe some of the steps we are taking to ensure that your special medical needs are met."[19]

Approximately one month later, Kaiser also gave these patients access to a Web site to provide them with updates about the transition in kidney transplant care and to inform them about their continued care as Kaiser members, said Schiffgen. Kaiser patients and their families are the primary audience of the HMO's communication efforts, he added, and these communication efforts were designed to reassure them that they "have not been left in limbo."

Perhaps just as important as continually communicating with patients, Kaiser also took steps to express ongoing good intentions to Kaiser Permanente doctors and staff, said Schiffgen. Frequent memos were sent to staff in advance of press briefings. These messages were meant to convey that Kaiser's Northern California kidney transplant program would continue to provide its patients with pre- and

CHAPTER FOUR

post-transplant care, "and that's something [physicians and employees] should feel good about," he said.

Without interviewing a significant number of affected Kaiser employees, it is difficult to know how they felt about the scandal and how the organization's communication efforts influenced them. At a minimum, it was a wise decision to offer organization members counter-narratives to the attacks against Kaiser's image.

Conclusion

In terms of its implementation of communication strategy, Kaiser Permanente's response to the scandal involving its Northern California kidney transplant program was not effective. This is not meant as a criticism of the well-meaning communication professionals at Kaiser Permanente. Moreover, it is likely that in the context of this issue Kaiser's corporate culture created an undisciplined communication response by several officials within the organization. Kaiser's response to attacks helped to create three communication-related threats that worked against its efforts to restore its image.

First, Kaiser officials—particularly clinicians—spoke candidly to the press without making sure that they could substantiate their claims that Kaiser did not harm patients. Some of these statements do not take into consideration the perceptions of confused and anxious patients.

Second, Kaiser officials made statements to the press that were not accurate. These statements served to further damage Kaiser's image because they raised questions about whether the organization was responding ethically to the attacks on its image.

Third, in the face of accusations that it ignored patient complaints and questions, Kaiser took too long to deliver a compelling counter-narrative to the patients on its kidney transplant waiting list.

Implications for action

This case study shows how quickly rhetorical attacks against a healthcare organization can develop. When charged with committing acts that have harmed patients, healthcare organizations face particularly difficult situations. By analyzing Kaiser Permanente's attempts to defend its public image, healthcare executives can think about how their organizations might respond to charges of administrative and clinical errors. Below are a few key lessons to take away from Kaiser's experience.

Craft credible statements

Using more image restoration strategies does not necessarily make for a stronger communication effort. One finely crafted, accurate strategy can be much more persuasive than multiple, disconnected, and contradictory messages across all of the broad categories of image restoration. Benoit points out that with "any persuasive message, image restoration attempts are unlikely to succeed when the evidence available to the audience contradicts those claims."[20] Statements by Kaiser officials were contradicted by evidence from its attackers, and Kaiser acknowledged to the *Times* that it provided inaccurate statements to reporters. Healthcare organizations under attack would be well advised to consider carefully attempts to deny and minimize claims of patient harm. If evidence later backs up the claim that patients were harmed by the organization, the perceived offensiveness is twofold: The public learns about the harm to patients, and the public believes that the organization conspired to cover it up.

Decide on the story and storytellers

To deal effectively with a crisis event, create an accurate yet persuasive story and make sure that only approved officials in your organization tell it to the press. In healthcare, with so many expert staff members on the frontlines and easily accessible, it can be difficult to create a single counter-narrative that effectively addresses your attackers' charges. Note well how journalists exploited this weakness at Kaiser. They quoted a number of Kaiser representatives, and it is likely that many of these executives and clinicians did not have training in dealing with the press. The safest method for media relations is to designate a point person within your organization to respond to the media, and communicate to all other physicians and staff that media inquiries should be directed to your press office.

Bolster with caution

When dealing with a crisis event, it is safe to assume that all statements and actions will be thoroughly scrutinized. Officials in your organization will want to counter attacks by boasting of your good deeds. Make sure these are on point and accurate. In Kaiser's case, boasting about its kidney transplant success rate was likely a simple error in which the official did not have accurate data when responding to the media, and Kaiser did the right thing by correcting this information with the press. However, Kaiser should have known that this claim of success would bring scrutiny about its kidney selection process.

Communicate quickly and directly to stakeholders

Don't count on your statements to the press to reach stakeholders. Media gatekeepers will give at least an equal amount of credibility and time to your attackers. Therefore, create a hierarchy of your stakeholder groups, and communicate with them directly. In this case study, Kaiser Permanente seemed unprepared for the number of attacks it received, and it did not communicate directly with all 2,000 patients on its kidney transplant waiting list until it sent letters some 20 days after initial news reports. This represents a long exposure of a salient audience to critics' messages. This case study highlights the importance for healthcare organizations to create strategic responses as quickly as possible to enhance their influence on stakeholders.

Offer a sincere apology

Healthcare organizations are urged to offer an apology when they determine that salient stakeholders' perceptions of the offense and the organization's responsibility have considerable merit. An apology can be a symbolic conclusion by the organization that allows it to then take corrective action and put the offensive act behind it. Although Kaiser's Thode offered a carefully worded apology that appeared to be sincere, it came after many puzzling and largely ineffective statements by Kaiser officials to the press. This creates the perception that the apology was made after other restorative strategies did not work. Furthermore, Thode's minimization efforts, as reported in the *Times*, further weakened the apology's influence.

Chapter Four

> *Give them something to feel good about*
>
> It is widely recognized that healthcare organizations have good intentions—especially in the context of patient care. In addition, combining a thoughtful apology with a statement of good intentions would be particularly effective for healthcare providers who are under attack. Kaiser's good intentions strategies would have been much more influential if they did not compete with the plethora of less-than-successful image restoration strategies, especially those that were later refuted. However, unlike many other Kaiser strategies, its statements of good intentions did not work against it. By and large, including statements about your healthcare organization's commitment to the community and affected patients is well advised.

Endnotes

1. Charles Ornstein and Tracy Weber, "Kaiser Put Kidney Patients at Risk," *The Los Angeles Times* (May 3, 2006): A1.

2. Charles Ornstein and Tracy Weber, "State Steps In on Kaiser Transplants," *The Los Angeles Times* (May 10, 2006): A1.

3. Tracy Weber and Charles Ornstein, "Kaiser Transplant Patients Express Their Fear and Fury," *The Los Angeles Times* (May 6, 2006): A1.

4. Anna Werner, "Kaiser Permanente SF Facing Transplant Troubles," CBS 5, *http://cbs5.com/investigates/local_story_122224454.html* (accessed July 12, 2006).

5. Tracy Weber and Charles Ornstein, "Kaiser Denied Transplants of Ideally Matched Kidneys," *The Los Angeles Times* (May 4, 2006): A1.

6. Tracy Weber and Charles Ornstein, "Kaiser Official Apologizes," *The Los Angeles Times* (May 11, 2006): A1.

7. *Los Angeles Times* Editorial, "Trust and Transplants," *The Los Angeles Times* (May 12, 2006): B12.

8. Tracy Weber and Charles Ornstein, "Kaiser Halts Kidney Venture,"*The Los Angeles Times* (May 13, 2006): A1.

9. Charles Ornstein and Tracy Weber, "Kaiser Told to Fix Problems with Kidney Patient Hotline," *The Los Angeles Times* (May 18, 2006): B1.

10. Charles Ornstein and Tracy Weber, "U.S. Berates Kaiser Over Kidney Effort," *The Los Angeles Times* (June 24, 2006): B1.

11. Tracy Weber and Charles Ornstein, "Kaiser Slow to Move Patients," *The Los Angeles Times* (June 16, 2006): B3.

12. William Benoit, *Accounts, Excuses, and Apologies: A Theory of Image Restoration Strategies* (New York: State University Press, 1995), 71.

13. Anna Werner, "Kaiser Apologizes for Transplant Troubles," CBS 5, http://cbs5.com/investigates/local_story_131211648.html (accessed July 13, 2006).

14. Charles Ornstein and Tracy Weber, "Kaiser Slow to Transfer Patients," *The Los Angeles Times* (May 5, 2006): A1.

15. Kaiser Permanente, "Diane Gage Lofgren Named Senior Vice President, Brand Strategy, Communications and Public Relations for Kaiser Foundation Health Plan, Inc., and Kaiser Foundation Hospitals," Kaiser Permanente, *ckp.kp.org/newsroom/national/archive/nat_060623_dianegagelofgren.html* (accessed: July 10, 2006).

16. U.S. News & World Report, "Kaiser Foundation Health Plans—Northern California (HMO)," U.S. News & World Report, *www.usnews.com/usnews/health/best-health-insurance/directory/dir_133_1530.htm* (accessed: July 13, 2006).

17. Chris Rauber, "Kaiser Names New Brand Strategy Exec," San Francisco Business Times, *http://sanfrancisco.bizjournals.com/sanfrancisco/stories/ 2006/06/19/daily52.html?t=printable* (accessed: July 6, 2006).

18. Victoria Colliver, "State to Investigate Kaiser Organ Program," *The San Francisco Chronicle* (May 4, 2006): C1.

19. R. Michael Alexander and Bruce Blumberg, M.D., "Dear {Patient First Name}," CBS 5, *http://cbs5.com/investigates/local_file_145193550* (accessed: July 5, 2006).

20. William Benoit, *Accounts, Excuses, and Apologies: A Theory of Image Restoration Strategies* (New York: State University Press, 1995), 130.

5

Loyola University Medical Center responds after infant abduction

Alarms sounded at Loyola University Medical Center, alerting staff that a baby had been taken from the nursery. During the commotion, a young woman exited the hospital. Under her coat, she hid a four-pound infant boy. Nearly nine hours later, he was found dead in the woman's apartment. Among the many questions this tragedy raised, one stood out: How could Loyola let this happen?

CHAPTER 5

Loyola University Medical Center responds after infant abduction

About Loyola University Medical Center

Loyola University Medical Center serves as the hub of Loyola University Health System, a private Jesuit Catholic provider based in Chicago's western suburbs. It includes Loyola University Hospital, a 523-bed facility with a Level 1 trauma center and a burn center. According to its Web site, Loyola University Health System employs about 6,800 and has 1,324 faculty members, 458 residents, 117 fellows, and 552 medical students. The medical center's profits rose 11% in 2005 to $12.7 million, from $11.4 million in 2004, and revenue increased nearly 6% in the same year, to $694.4 million from $656.7 million, according to a *Chicago Tribune* report.[1]

CHAPTER FIVE

Background of the controversy

The first headcount of babies at Loyola University Medical Center's nursery found that none were missing on May 1, 2000, but a double-check by staff moments later discovered 13-day-old Zquan Wakefield was gone.[2] This was not supposed to happen. When the baby's electronic bracelet tripped the alarms, the exits should have locked automatically. Something went very wrong.

The medical center's security searched the 61-acre campus for more than an hour but then called the Cook County Sheriff's Department to report that the baby was missing. After police viewed the hospital's security video, they brought Vanecha Cooper and her boyfriend in for questioning. Cooper had been to Loyola that evening to visit another family at the nursery. Police found inconsistencies in her story and returned to Cooper's apartment for a search.[3]

By this time, several hours had passed since Zquan disappeared. Police discovered his tiny body facedown in a laundry hamper. The autopsy revealed that the premature baby died of asphyxiation.[4] Apparently, Cooper attempted to hide Zquan from police during their first visit. He likely died within a half-hour of being left there.

In some ways, Cooper fit the classic description of baby abductors, according to experts. She had a history of miscarriages and had lied to friends and family that she was pregnant.[3] In one important way, this case broke the mold; most of the time, babies kidnapped from hospitals are returned unharmed.

Was the hospital responsible?

This crime sparked outrage, but not only against the abductor. The press and lawyers for Zquan's family wanted to know how someone unrelated to the child could scoop him up from the nursery, slip past security, and then walk off into the night. "We are fairly certain that, in this case, the hospital had some security lapses which directly resulted in this kidnapping," said James Montgomery, an attorney for the Wakefield family.[4] He suggested that Cooper had visiting privileges for another child, but that hospital staff should have kept her away from Zquan's crib.

The *Chicago Sun-Times* editorial staff criticized the hospital for waiting an hour to call police.[5] A former Loyola maternity nurse told *Chicago Tribune* writers that the hospital's security was "disorganized" and that she was not surprised that a hospital guest got into the nursery.[3]

After looking into the incident, Loyola officials explained that faulty wiring caused a malfunction. A door was supposed to close and lock the moment the baby's bracelet set off alarms. Instead, Cooper was able to escape, concealing Zquan under her coat.[2]

Not human error

Trisha Cassidy, senior vice president for health system development and strategy for Loyola Health System, said, "This is not the result of human error. This was a technological error. We put technological systems in place so we don't have to rely on human judgment. That system failed."[2] She also acknowledged that staff initially counted the number of babies incorrectly, and she said hospital staff waited an hour to call police because Loyola security guards were better suited to conduct the initial search of the campus.[2]

Chapter Five

In an editorial, the *Chicago Tribune* criticized Cassidy's statements: "No electronic gizmo can substitute for an alert staff who know their charges by name and know who should, and shouldn't, be hovering over each bassinet."[6]

The Centers for Medicare & Medicaid Services threatened to cut off funding, but later federal regulators accepted Loyola's plan to prevent future abductions. Regulators cited several security problems that might have contributed to the kidnapping. For instance, they said that Loyola did not have a "specific schedule" for testing its alarm system.[7] In the end, inspectors said the actions included in Loyola's plan to prevent future abductions were appropriate. Loyola also created new policies and trained staff.

Loyola makes changes

In a prepared statement, Loyola University Health System's President and CEO Dr. Anthony L. Barbato said, "We have undertaken a comprehensive review of hospital policies and procedures related to our maternity and newborn nursery and have implemented changes that reflect what we have learned from this event."[8]

Separate lawsuits were filed against Loyola on behalf of the baby's mother and father. In August 2000, Loyola Health System announced that it had reached settlement agreements with both parents. Loyola did not share the dollar figure of the settlements with the public. In a prepared statement, it said that it accepted "a measure of responsibility" for the incident.[9] "Loyola Health System recognizes that no amount of money could compensate the Wakefield family for their tremendous loss," the system said in the statement. "We have settled the separate lawsuits filed by the mother and the man identified as the father. We hope our settlement will provide the family with some sense of closure to this terrible tragedy."[9]

Loyola University Medical Center's stakeholders

As Loyola responded to attacks on its public image, rhetoric by the following individuals and groups shaped the public's perceptions:

- Loyola officials
- Federal regulators
- Law enforcement
- Lawyers for the Wakefield family
- Staff
- A former employee
- Loyola media affairs office
- Media

Loyola University Medical Center communicated with many of the above stakeholders. It conducted an investigation and worked to enhance policy and training for its maternity and nursery based on recommendations from regulators. Officials provided information to the media and created press statements. Loyola also reached a settlement agreement with lawyers representing the Wakefield family.

Although a federal regulator said Loyola was "very forthright in acknowledging the problems once they got our statement of deficiencies,"[8] medical center officials did not publicly take accountability for any problems other than the security door's faulty wiring.

Chapter Five

Analysis of efforts to restore Loyola University Medical Center's public image

Zquan's kidnapping was the first infant abduction in Loyola's 30-year history. The vast majority of infant abduction attempts involving hospitals fail—typically, the kidnapper never makes it outside of the hospital. Rarer still are abductions from hospitals that end in an infant's death.

Certainly, Loyola officials had a host of concerns related to this incident. To name a few, they needed to investigate to find a root cause for potential security failures, plan for the inevitable lawsuit from the baby's family, and respond appropriately to federal investigators. In terms of Loyola's public image, hospital officials attempted these restoration strategies: reducing the offensiveness of the event through differentiation and compensation, and corrective action.

Differentiation

As one variable of reducing the offensiveness of an act, an organization might attempt to engage in differentiation. As Benoit points out, "Here the rhetor attempts to distinguish the act performed from other similar but less desirable actions."[10] Note well how Loyola's Cassidy attempted to put the onus on a technological malfunction, differentiating this from an offense committed by Loyola staff. "This is not the result of human error. This was a technological error," she said. "We put technological systems in place so we don't have to rely on human judgment. That system failed."[2]

Considering that regulators criticized Loyola for not properly testing this technology and that Cassidy herself acknowledged that staff miscounted the number of infants and did not immediately call police, this attempt to evade responsibility was not successful. Indeed, Cassidy's statement became fodder for the *Tribune's* editorial writers.

Compensation

Compensation is another strategy that can be used to reduce the offensiveness of an act. "If the accuser accepts the proffered inducement, and if it has sufficient value, the negative affect from the undesirable act may be outweighed, restoring reputation," notes Benoit.[11] In this case, Loyola reached a legal, monetary settlement with lawyers for the Wakefield family. As both sides rightly pointed out, money cannot adequately compensate for the lost life of an infant child. For this reason, Loyola's compensation strategy is not particularly effective at restoring its public image, although it did bring closure to its legal exposure.

Corrective action

Although Loyola officials did not admit that errors by staff facilitated Zquan's abduction, the medical center took a number of corrective measures in response to deficiencies outlined by regulators. "The hospital failed to ensure that systems designed to ensure the safety of patients in the newborn nursery were tested, functioning adequately, and implemented in the event of the abduction of an infant," stated the report from the Health Care Financing Administration.[6]

Loyola officials apparently satisfied this highly important stakeholder organization, which had the power to withhold hundreds of millions of dollars in Medicare and Medicaid funding.[6] "We're very pleased they took action to come back into compliance," said HCFA's Bob Daly.[8]

Chapter Five

As far as its relationship with Medicare and Medicaid was concerned, Loyola's corrective action strategy worked well. However, in terms of its public image, this corrective action strategy conflicted with statements from Loyola officials that the organization was responsible for only the technical error of faulty wiring at the medical center. This disconnect between words and action leaves one with the feeling that Loyola officials were less than forthright in their public statements.

Conclusion

Perhaps Loyola's communication strategy was hindered by the threat of lawsuits and related settlement discussions. Sometimes when organizations are concerned about legal exposures, they restrict public communication and shy away from taking accountability. However, this is the rhetoric of damage control and not crisis communication. Comments by Loyola officials to the media were not effective at restoring its public image. Moreover, they lacked credibility by conflicting with the statements from federal regulators.

Implications for action

Quite possibly the worst case scenario for a healthcare organization, this case study shows how one organization responded to an infant abduction and death. This is a damaging situation no matter what the organization's rhetorical response. Here are a few suggestions for attempting to implement an effective restorative strategy:

Take a hard look at yourself
When leaders discover that a terrible event has occurred, they might be

tempted to mount a defense that attempts to puts the organization in the best possible light. This is not always the role of communication. Consider carefully the plausibility and fidelity of statements that reject accountability for offensive acts. For example, blaming an offense on a technological system that your organization's staff is supposed to test and maintain is not likely to reduce the offensiveness of the perceived wrongdoing.

Offer regrets

In a statement of regret, your organization does not need to accept responsibility for the perceived offense. When the vulnerable are victimized, a statement of regret offered early and sincerely can generate good will toward the organization and its senior leaders. In addition, this statement should be attributed to a senior official. In announcing the settlement agreement with the Wakefield family, the prepared statement was not attributed to a Loyola official, and the statement read more like a legal document than it did a sincere showing of regret.

Make corrections quickly

According to regulators, Loyola responded promptly to comply with Medicare and Medicaid. The organization created new policies, trained staff, and corrected technical problems. In addition, Loyola officials communicated these changes effectively to HCFA inspectors.

CHAPTER FIVE

Endnotes

1. Bruce Japsen, "Loyola Health System CEO Quits," *Chicago Tribune* (June 15, 2006): Business, 3.

2. Jeremy Manier and Jana Hanna, "Baby Alive When Hidden From Police," *Chicago Tribune* (May 4, 2000): News, 1.

3. Jeremy Manier and Vanessa Gezari, "Baby Dies in Hospital Kidnap," *Chicago Tribune* (May 3, 2000): News, 1.

4. Rick Hepp, "Lawyer: Family of Abducted Infant Will Sue," *Chicago Tribune* (May 4, 2000): Metro, 1.

5. *Chicago Sun-Times* Editorial, "Anatomy of a Tragedy," *Chicago Sun-Times* (May 4, 2000): Editorial, 35.

6. *Chicago Tribune* Editorial, "Loyola's Painful Lesson on Safety," *Chicago Tribune* (May 4, 2000): Editorial, 30.

7. Jeremy Manier, "Loyola Hospital May Avoid Penalty," *Chicago Tribune* (May 17, 2000): Metro, 1.

8. Jeremy Manier, "Loyola No Longer Faces Loss of Medicare Funds," *Chicago Tribune* (May 26, 2000): Tribune West, 10.

9. Vanessa Gezari, "Settlement in Hospital Kidnapping," *Chicago Tribune* (August 18, 2000): Metro, 1.

10. William Benoit, *Accounts, Excuses, and Apologies: A Theory of Image Restoration Strategies* (New York: State University Press, 1995), 77.

11. William Benoit, *Accounts, Excuses, and Apologies: A Theory of Image Restoration Strategies* (New York: State University Press, 1995), 78.

6

Hacker accesses patient information at University of Washington Medical Center

A computer hacker announced to the press that he was able to use the Internet to infiltrate the University of Washington Medical Center's network. He stole approximately 5,000 administrative records, which included sensitive patient data. This information security breach put the scrutiny of the national media on UWMC officials and caused some patients to wonder whether their medical records were safe.

CHAPTER 6

Hacker accesses patient information at University of Washington Medical Center

About the University of Washington Medical Center

Known as a major teaching hospital, the University of Washington Medical Center in Seattle is a 450-bed comprehensive medical care facility. In its 2002 annual report, UWMC boasted a score of 94 out of a possible 100 from the Joint Commission on Accreditation of Healthcare Organizations. Nearly 400 attending physicians are full-time faculty members at the UWMC's School of Medicine. Of the 176 medical centers ranked in the 2006 edition of *U.S. New & World Report's* America's best hospitals list, UWMC earned the 10th spot. For fiscal year 2002, UWMC reported an operating income of about $8.5 million from approximately $444.5 million of total revenue.

Chapter Six

Background of the controversy

A man claiming to be from the Netherlands and calling himself "Kane" told an online reporter that he successfully hacked into the University of Washington Medical Center's network and stole about 5,000 patient records.[1] The revelation, published in December 2000, set off a wave of concern and speculation about the security of UWMC's medical records system.

The electronic break-in occurred months prior to news reports. At the time, the medical center's chief information officer suspected that there was a network infiltration and took steps to limit the hacker's access. He was unaware that any patient data had been accessed.[1]

Confronted with evidence

When newspaper reporters asked UWMC officials about the hack, they strongly denied that patient information had been stolen; however, reporters later confronted them with images of patient files that were provided by the hacker.

After an investigation, UWMC concluded that the hacker used the Internet to filch thousands of administrative files containing patient names, conditions, home addresses, and Social Security numbers.[3] This news made patients feel insecure about their personal health information stored at the prominent medical center.

"Sometimes the consequences of having your medical records revealed could cost you a great deal," said Edwin Gould, a heart patient at UWMC whose data had been stolen.[3]

"I don't think anything is secure anymore," said John Carter, another patient whose records were pilfered.[3]

UWMC shares information

Newspapers reported that the lost Social Security numbers put patients at risk for identity theft, and medical information could be used by employers or insurers to discriminate against people.[2] There was no evidence of the hacker using the data in these ways. In fact, he claimed to be an "ethical hacker" who aimed to blow the whistle on security lapses at the major medical center.[3]

UWMC took steps to improve network security and informed some 5,000 patients about the breach. Hospital officials set up a hotline for worried patients and called the FBI to investigate. UWMC also quickly called a press conference to inform the public about its initial findings.

"The first thing I would say is how sorry we are that this happened," said Dr. Eric Larson, UWMC's medical director. "We take this incredibly seriously."[2]

An important distinction

Medical center officials worried that there was a public perception that UWMC's medical records system was penetrated. They used the news conference to point out that patient data were taken as part of stolen administrative files used for things like scheduling. "It is important to emphasize that there is no evidence that anyone has breached our main electronic medical-record system," said Tom Martin, CIO for UWMC. "While the initial report said that the hacker had control of major potions of our database, we have learned that he only got into specific administrative databases."[4]

Chapter Six

However, that the hacker made off with administrative records including patient data—not highly confidential medical records—is a distinction lost on people exposed to news sound bites, said Larson.[5]

Martin was quite frank during his discussions with reporters about the information security lapses at UWMC. According to Martin, the hacker used a Web site in the pathology department to break into the network. From there, the hacker's computer program captured legitimate passwords and user names that allowed him to access the administrative records.[3]

UWMC officials said that they had fortified security and promised to be more vigilant. "We want to let [patients] know how seriously we take this matter and that we are responding in the most responsible manner possible," said Martin.[4]

Hacker accesses patient information at University of Washington Medical Center

University of Washington Medical Center's stakeholders

These groups and individuals created the meaning of UWMC's public image as the medical center worked to restore its reputation:

- Senior leaders
- The chief information officer
- PR staff
- Patients
- Media
- FBI
- Kane (the hacker)

When stories about the stolen records were published, UWMC responded by holding a news conference, contacting affected patients, setting up a telephone hotline, and creating a Web page resource. Officials also alerted the FBI.

Analysis of efforts to restore the University of Washington Medical Center's public image

The University of Washington Medical Center attempted multiple strategies of image restoration discourse. The organization's leaders attempted to reassure the public about the security of their medical information and set some inaccuracies straight. Below, I show how UWMC attempted to implement these strategies: denial, reducing offensiveness of the event by differentiation and minimization, corrective action, and mortification.

CHAPTER SIX

Denial
After a story about the hack ran on the Web site SecurityFocus.com, UWMC officials called the report "completely inaccurate." A report in the *Seattle Post-Intelligencer* also stated that medical center officials strongly denied the report.[2] However, when the media shared images of some of the patient records stolen from UWMC's system, officials had to change their position.

When an organization believes that a damaging claim against it is inaccurate, there's an initial instinct to take a defensive posture and deny the claim's validity. This is not recommended unless the organization has completed an exhaustive review of the situation and is certain of the accuracy of its denial. The initial response from an organization in crisis can set the tone of its relationship with the media throughout a crisis. Starting off by providing media members with inaccurate or misleading information can have a damaging effect on the organization's credibility throughout the duration of the crisis.

Although it was unwise of UWMC officials to deny the claims of a security breach outright, it seems to have been a simple error that did not cause the organization to lose credibility with the press.

Differentiation
UWMC was extremely aware that patients might worry that their medical records were stolen. Larson rightly pointed out that the public might not be able to distinguish the fact that the hacker downloaded copies of administrative records, which included some patient information, but that he did not gain access to the highly confidential patient medical records. Therefore, the CIO of the medical center shared detailed information about the electronic intrusion with the media in an attempt to influence how the story would be covered. "There is no evidence that

anyone has breached our main electronic medical records," said the hospital in a prepared statement. "We assure patients and the public that this system remains fully protected by the highest levels of security possible."[1]

UWMC's differentiation strategy worked well. In several reports, newspapers pointed out the difference between medical records and administrative records. As *ComputerWorld* noted, "In a statement, the hospital said the copied information wasn't directly related to the delivery of care to its patients. Rather, the information was stored in administrative databases and was used for patient tracking and following up on research studies."[1]

Minimization

In addition to making the distinction between the types of files the hacker stole, UWMC attempted to minimize the effect of the event. By suggesting that computer hackers have broken into other organizations, UWMC tried to minimize the perceived offense that its system could allow such an intrusion. UWMC's Martin suggested that even Microsoft had been the victim of computer hacking earlier that year. "And they're experts," he said. "And we're working with experts here. It's just a very unfriendly environment we're working in, and we're constantly being vigilant."[2]

Pointing out that Microsoft had been hacked does not do any harm to the overall effectiveness of UWMC's image restoration strategy. For certain, with new technology come new abuses of the technology. However, this is not a particularly effective argument. The public holds healthcare organizations to a much higher standard when it comes to protecting patient confidentiality than the standard to which it holds Microsoft.

Chapter Six

Corrective action

In perhaps its strongest effort to restore its public image, UWMC officials clearly explained in detail that they understood how the hacker succeeded in copying patient information. Officials said that they took steps to improve the organization's network security and moved all patient information behind a firewall. In addition, UWMC said it would educate staff and patients about information security.

"The incident has been a galvanizing event for the hospital and the university as a whole," said Martin. "Lately, I do two security presentations a week to senior decision-makers."[5]

Mortification

Officials at UWMC made a public apology for the institution's lack of information security safeguards, and the CIO attempted to improve the public's understanding of the incident by sharing with the press details of the hack and the organization's plans to correct network exposures.

"The first thing I would say is how sorry we are that this happened," said Larson at the beginning of a press conference about the hacking episode. This apology, coupled with Martin's transparency, helped show stakeholders that UWMC would take better precautions with patient information on its computer network.

Conclusion

Overall, officials at the University of Washington Medical Center implemented image restoration strategies effectively. Once the *Washington Post* provided conclusive evidence of the security breach, officials displayed openness about their problems and offered credible statements about limiting future lapses. In addition, they were able to differentiate administrative records from medical records well enough for the media to report about the incident accurately.

Implications for action

With HIPAA privacy rules in effect, the risk of allowing patient information to fall into the wrong hands is even greater today than it was for UWMC at the time of this incident. However, there are many things to learn from the medical center's response. Here are some key lessons from UWMC's experience:

Check your facts
Sometimes others have better information about your crisis situation than you. Choose the words of your denial carefully so that you do not contradict yourself later. In UWMC's case, reporters said officials "strongly denied" accounts of stolen patient data. No doubt officials believed their statements to be true, but the reality was that they did not know whether they had a security breach. Better to be honest and say that you are conducting a detailed investigation that you hope will provide reliable information.

Open the curtains and let in the light
When there is an obvious problem—such as an information security break-in—stakeholders want to know that you have the solution. One strong way to build back trust is to share information about what went wrong and how you are going to correct it. UWMC's Martin acknowledged his network's shortcomings and explained to the press how he was going to fix weaknesses in the system.

Chapter Six

> ### *Help the press get the facts right*
> Although some might not appreciate the differences, administrative records are not medical records. UWMC made a point of explaining the importance of the difference to journalists, and many press reports made the distinction. If you see an opportunity for public confusion, make the facts clear—especially when they work out to your favor.

Endnotes

1. Marc L. Songini, "Hospital Confirms Hacker Stole 5,000 Patient Files," *ComputerWorld* (December 18, 2000): 7.

2. Chris Grygiel, "FBI Joins UW in Hacking Investigation," *Seattle Post-Intelligencer* (December 9, 2000): A11.

3. Robert O'Harrow, Jr., "Hacker Accesses Patient Records," *Washington Post* (December 9, 2000): E01.

4. Mike Cane, "UWMC Hacker Incident Prompts Security Increase," *The University of Washington Daily* (online), *archives.thedaily.washington.edu/2001/010401/N4.UWHacker.html* (accessed July 27, 2006).

5. Susan Rudi Rodsjo, "Coping with Insecurities," HealthLeaders Media, *www.healthleadersmedia.com/print.cfm?content_id=26080&parent=107* (accessed July 19, 2006).